Introduction to Modern
System Sciences

MW00845664

Dragos B. Chirila · Gerrit Lohmann

Introduction to Modern Fortran for the Earth System Sciences

 Springer

Dragos B. Chirila
Gerrit Lohmann
Climate Sciences, Paleo-climate Dynamics
Alfred-Wegener-Institute
Bremerhaven
Germany

ISBN 978-3-662-51082-7   ISBN 978-3-642-37009-0   (eBook)
DOI 10.1007/978-3-642-37009-0

Springer Heidelberg New York Dordrecht London

Printed on acid-free paper

Springer is part of Springer Science+Business Media (www.springer.com)

*We dedicate this text to the contributors
(too numerous to acknowledge individually)
of the free- and open-source-software
community, who created the tools that
enabled our work.*

Preface

Since the beginning of the computing age, researchers have been developing numerical Earth system models. Such tools, which are now used for the study of climate dynamics on decadal- to multi-millennial timescales, provide a virtual laboratory for the numerical simulation of past, present, and future climate transitions and ecosystems. In a way, the models bridge the gap between theoretical science (where simplifications are necessary to make the equations tractable) and the experimental science (where the full complexity of nature manifests itself, as multiple phenomena often interact in nonlinear ways, to form the final signal measured by the apparatus). Models provide intermediate subdivisions between these two extremes, allowing the scientist to choose a level of detail that (ideally) strikes a balance between accuracy and computational effort.

The development of models has accelerated in the last 50 years, largely due to decreasing costs of computing hardware and emergence of programming languages accessible to the non-specialist. Fortran, in particular, was the first such language targeting scientists and engineers, therefore it is not surprising that many models were written using this technology. To many, however, this long history also causes Fortran to be associated with the punched cards of yesteryear and obsolete software practices (hence the quotation above). A programming language, however, evolves to meet the demands of its community, and such was also the case with Fortran: object-oriented and generic programming, a rich array language, standardized interoperability with the C-language, free-format (!), and many more features are now available to Fortran programmers who are willing to take notice.

Unfortunately, many of the newer features and software engineering practices that we consider important are only discussed in advanced books or in specialized reference documentation. We believe this unnecessarily limits (or delays) the exposure of beginning scientific programmers to tools, which were ultimately designed to make their work more manageable. This observation motivated us to write the present book, which provides a short "getting started" guide to modern Fortran, hopefully useful to newcomers to the field of numerical computing within *Earth system science* (ESS) (although we believe that the discussion and code examples can also be followed by practitioners from other fields). At the same time, we hope that readers familiar with other programming languages (or with earlier revisions of the Fortran-standard) will find here useful answers for the "How do I do X in modern Fortran?" types of questions.

Chapters Outline

In Chap. 1, we start with a brief history of Fortran, and succinctly describe the basic tools necessary for working with this book. In Chap. 2, we expose the fundamental elements of programming in Fortran (variables, I/O, flow-control constructs, the Fortran array language, and some useful intrinsic procedures). In Chap. 3, we discuss the two main approaches supported by modern Fortran for structuring code: *structured programming* (SP) and *object-oriented programming* (OOP). The latter in particular is a relative newcomer in the Fortran world.

The example-programs (of which there are many in the book) accompanying the first three chapters are intentionally simple (but hopefully still not completely uninteresting), to avoid obfuscating the basic language elements. After practicing with these, the reader should be well equipped to follow Chap. 4, where we illustrate how the techniques from the previous chapters may be used for writing more complex applications. Although restricted to elementary numerical methods, the case studies therein should resemble more of what can be encountered in actual ESS models.

Finally, in Chap. 5 we present additional techniques, which are especially relevant in ESS. Some of these (e.g., namelists, interoperability with C, interacting with the *operating system* (OS)) are Fortran features. Other topics (I/O with *NETwork Common Data Format* (netCDF), shared-memory parallelization, build systems, etc.) are outside the scope of the Fortran language-standard, but nonetheless essential to any Fortran programmer (the netCDF is ESS-specific).

Language-Standards Covered

The core of the book is based on Fortran 95.[1] Building upon this basis, we also introduce many newer additions (from Fortran 2003 and Fortran 2008[2]), which complete the discussion or are simply "too good to miss"—for example OOP,

---

[1] This was, at the time of writing, the most recent version with ubiquitous compiler support.

[2] Many compilers nowadays have complete or nearly complete support for these newer language-standard revisions.

interoperability with the C-language, OS integration, newer refinements to the Fortran array language, etc.

Disclaimers

- Given the wide range of topics covered and the aim to keep our text brief, it is obvious that we cannot claim to be comprehensive. Indeed, good monographs exist for many topics, which we only superficially mention (many further references are cited in this text).
- Finally, we often provide advice related to what we consider good software practices. This selection is, of course, subjective, and influenced by our background and experiences. Specific project conventions may require the reader to adapt/ignore some of our recommendations.

How to Use this Book

Being primarily a compact guide to modern Fortran for beginners, this book is intended to be read from start to finish. However, one cannot learn to program effectively in a new language just by reading a text—as in any other "craft", practice is the best way to improve. In programming, this implies reading and writing/testing as much code as possible. We hope the reader will start applying this philosophy while reading this book, by typing, compiling, and extending the code samples provided.[3]

Readers with programming experience may also use "random access," to select the topics that interest them most—the chapters are largely independent, with the exception of Chap. 5, where several techniques are demonstrated by extending examples from Chap. 4.

Due to the "breadth" of the book, many technical aspects are covered only superficially. To keep the main text brief, we opted to provide as footnotes suggestions for further exploration. Unfortunately, this led to a significant number of footnotes at times; the reader is encouraged to ignore these, at least during a first reading, if they prove to be a distraction.

---

[3] Nonetheless, the programs are also available for download from SpringerLink. The authors also provide a code repository on GitHub: assuming a working installation of the git version-control system is available, the code repository can be "cloned" with the command:

```
git clone https://github.com/dchirila/imf_ess.git
```
.

Acknowledgments

The idea of writing this book crystallized in the spring of 2012. Almost 2 years later, we have the final manuscript in front of us. Contributions from many people were essential during this period. They all helped in various ways, through discussions about the book and related topics, requests for clarifications, ideas for topics to include, and corrections of our English and of other mistakes, greatly improving the end result. In particular, we acknowledge the help of many (past and present) colleagues from the *Climate Sciences* division at *Alfred-Wegener-Institut, Helmholtz-Zentrum für Polar- und Meeresforschung* (AWI)—especially Manfred Mudelsee, Malte Thoma, Tilman Hesse, Veronika Emetc, Sebastian Hinck, Christian Stepanek, Dirk Barbi, Mathias van Caspel, Sergey Danilov, and Dmitry Sidorenko. We thank Stefanie Klebe for a very thorough reading of the final draft, which significantly improved the quality of the book.

In addition to our AWI colleagues, we received valuable feedback from Li-Shi Luo, Miguel A. Bermejo, and Dag Lohmann.

Our editors from Springer were very helpful during the writing of this book. In particular, we thank Marion Schneider, Johanna Schwarz, Carlo Schneider, Marcus Arul Johny, Ashok Arumairaj, Janet Sterritt, Agata Oelschlaeger, Dhanusha M. and Janani J. for kindly answering our questions and for their support.

Finally, we would like to thank our families and friends, who contributed with encouragement, support, and patience while we worked on this project.

Bremerhaven, Germany, May 2014   Dragos B. Chirila
                                                                 Gerrit Lohmann

Contents

Acronyms

ADE	Alternating-direction explicit
ADT	Abstract data type
API	Application programming interface
BC	Boundary condition
CA	Cellular automata
CAF	Co-array Fortran (http://www.co-array.org/)
CF	Climate forecast (http://cf-pcmdi.llnl.gov/documents/cf-conventions)
CFL	Courant-Friedrichs-Levy
CLI	Command line interface
CPU	Central processing unit
DAG	Directed acyclic graph
DSL	Domain-specific language
DT	Derived Data Type
EBM	Energy balance model
EBNF	Extended Backus-Naur form
ESS	Earth system science
FD	Finite differences
FE	Finite elements
FV	Finite volumes
GP	Generic programming
GPGPU	General-purpose graphics processing unit
GUI	Graphical user interface
HDD	Hard disk drive
HPC	High performance computing
HLL	High-level language
I/O	Input/output
IC	Initial condition
ID	Identifier
IDE	Integrated development environment

ILP   Instruction-level parallelism
LBM   Lattice Boltzmann method
LGCA   Lattice gas cellular automata
LHS   Left-hand side
MPI   Message Passing Interface
MRT   Multiple relaxation times
NAS   Network-attached storage
ODE   Ordinary differential equation
OOP   Object-oriented programming
OpenCL   Open Computing Language
OpenMP   Open MultiProcessing
OS   Operating system
PDE   Partial differential equation
PDF   Particle distribution function
PGAS   Partitioned Global Address Space (http://www.pgas.org/)
RAM   Random access memory
RB   Rayleigh-Bénard
RHS   Right-hand side
RNG   Random number generator
RPM   Revolutions per minute
SIMD   Single instruction, multiple data
SP   Structured programming
SPMD   Single program, multiple data
TDD   Test-driven development
TRT   Two relaxation times

Fortran Compilers

gfortran   GNU Fortran Compiler (see entry on gcc)
ifort   Intel Fortran Compiler® (http://software.intel.com/en-
            us/fortran-compilers)

Profiling Tools

gprof   GNU Profiler (part of binutils) (http://www.gnu.org/
            software/binutils/)
VTune   Intel VTune Amplifier XE 2013 (https://software.intel.
            com/en-us/intel-vtune-amplifier-xe)

Other Software Utilities

bash   Bourne-again shell (http://www.gnu.org/software/bash)
CMake   Cross Platform Make (http://www.cmake.org)

```
Cygwin   http://www.cygwin.com/index.html
gcc      GNU Compiler Collection (http://gcc.gnu.org)
ld       GNU linker (http://www.gnu.org/software/binutils)
gmake    GNU Make (http://www.gnu.org/software/make)
MinGW    Minimalist GNU for Windows (http://www.mingw.org)
SCons    Software Construction tool (http://www.scons.org)
```

Visualization/Post-processing Tools

CDO   Climate Data Operators (https://code.zmaw.de/projects/cdo)

GMT   Generic Mapping Tools (http://gmt.soest.hawaii.edu)

gnuplot   http://www.gnuplot.info

NCO   netCDF Operators (http://nco.sourceforge.net)

ParaView   Parallel Visualization Application (http://www.paraview.org)

Operating Systems

AIX   IBM Advanced Interactive eXecutive

Linux   GNU/Linux

OSX   Mac OS X®

Windows   Microsoft Windows®

Unix   Unix® (http://www.unix.org)

Text Editors

Emacs   GNU Emacs text editor (http://www.gnu.org/software/emacs)

gedit   Gedit text editor (http://projects.gnome.org/gedit)

joe   Joe's Own Editor (http://joe-editor.sourceforge.net)

Kate   Kate text editor (http://kate-editor.org)

Vim   Vim text editor (http://www.vim.org)

Software Libraries

ACML   Core Math Library (http://developer.amd.com/tools/cpu-development/amd-core-math-library-acml)

ATLAS   Automatically Tuned Linear Algebra Software (http://math-atlas.sourceforge.net)

BLAS   Basic Linear Algebra Subprograms

Boost.   http://www.boost.org/libs/program_options
Program_Options

ESSL   Engineering Scientific Subroutine Library

fruit	FORTRAN Unit Test Framework (`http://sourceforge.net/projects/fortranxunit`)
GAMS	Guide to Available Mathematical Software (`http://gams.nist.gov`)
HDF5	Hierarchical Data Format—Version 5 (`http://www.hdfgroup.org/HDF5/`)
JAPI	Java Application Programming Interface (`http://www.japi.de`)
LAPACK	Linear Algebra PACKage (`http://www.netlib.org/lapack`)
MKL	Intel® Math Kernel Library (`http://software.intel.com/en-us/intel-mkl`)
netCDF	NETwork Common Data Format (`http://www.unidata.ucar.edu`)
netlib	`http://www.netlib.org`
Winteracter	`http://www.winteracter.com`
Zenity	`https://help.gnome.org/users/zenity/stable`

Other Programming Languages

C	`http://en.wikipedia.org/wiki/C_(programming_language)`
C++	`http://en.wikipedia.org/wiki/C%2B%2B`
COBOL	Common Business Oriented Language
Java	`http://www.java.com/en/`
MATLAB	Matrix Laboratory® (`http://www.mathworks.com`)
octave	GNU Octave (`http://www.gnu.org/software/octave`)
Pascal	`http://en.wikipedia.org/wiki/Pascal_(programming_language)`
Python	`http://www.python.org`
R	The R Project for Statistical Computing (`http://r-project.org`)

Version Control Software

git	`http://www.git-scm.com`
mercurial	`http://mercurial.selenic.com`
monotone	`http://www.monotone.ca`
subversion	`http://subversion.apache.org`

Earth System Science Models

```
Planet          http://www.mi.uni-hamburg.de/Planet-
Simulator       Simul.216.0.html?&L=3
```

Organizations and Companies

AMD   Advanced Micro Devices Inc.
ANL   Argonne National Laboratory (`http://www.anl.gov`)
Apple   Apple Inc.
ASCII   American Standard Code for Information Interchange
AWI   Alfred-Wegener-Institut, Helmholtz-Zentrum für Polar- und Meeresfors-
        chung (`http://www.awi.de`)
GNU   GNU project—software project backed by the Free Software Foundation
        (FSF); the (recursive) acronym stands for GNU's Not Unix! (`http://`
        `www.gnu.org`)
IBM   International Business Machines Inc.
Intel   INTegrated ELectronics Inc.
OGC   Open Geospatial Consortium (`http://www.opengeospatial.`
        `org/standards/netcdf`)
UCAR   University Corporation for Atmospheric Research (`https://www2.`
        `ucar.edu`)
WMO   World Meteorological Organization (`http://www.wmo.int`)

Conventions in this Text

The following conventions are used for the code samples:

1. *Formatting and color scheme*
   Programs and code samples that would normally be typed in an editor, are
   shown in boxes, with the following conventions in place:

   - *keywords*[4]: dark gray, bold font
   - *character strings*: medium gray, normal font
   - *comments*: medium gray, italic font

2. *Code placeholders*

   - *Optional items* are emphasized using square brackets. When the reader wishes
     to include them within programs, the brackets should be removed.

---

[4] We choose to typeset Fortran keywords with lowercase letters, although the language is case-insensitive everywhere except inside character strings (so PROGRAM, program or PrOgRaM is all the same to the compiler).

- *Mandatory items* that should be supplied by readers, as well as *invisible characters* are emphasized using angle brackets, as in:

```
if( <logical expression> ) then
    print*, ''Expression was .true.''
end if

<Enter>
<Space>
```

It should be easy to infer from the context what these angle bracket expressions should be replaced with.

- *Combinations* of *optional* and *mandatory* items are sometimes highlighted by nesting of square and angle brackets, to distinguish the fact that including some items may unlock additional possible combinations.

3. With the exception of small snippets, *code listings* are accompanied by a caption, indicating the corresponding file in the source code tree available for download. Line numbers are only shown when they are specifically referenced in the text.

4. Where interaction with the Operating System (OS) is illustrated, we describe the process for the *GNU/Linux* (Linux) platform, using *Bourne-again shell* (bash), since this environment is easily accessible. Commands corresponding to such tasks are marked by a leading $-character (default shell-prompt in Linux); only the part after this marker should be typed.

5. Exercises are typeset on a dark-gray background, to distinguish them from the rest of the text.

6. Several notes appear as framed boxes on a light gray background.

7. *Naming conventions* It is usually considered good practice to adopt some rules for naming entities that are part of the program code. Although different developers may prefer a different set of such rules, it is generally a good idea to use a single convention *consistently* within a project, to reduce the effort required for understanding the code. Our particular conventions are explained below.

Naming Rules for Data

- Variables (both scalars or arrays) are named as things (nouns) or attributes (adjectives). When they consist of multiple words, camel-case is used, starting with a lowercase letter:

```
temperature , numIterations
```

- Variables that are part of a user-defined type follow the same rules as above, except that the first letter is always a lowercase "m":

```
mNx , mOutFilePrefix
```

- Constants are written in uppercase, and when they are composed of multiple words they are separated by underscores:

```
PI, MAX_NUM_ITERATIONS
```

- User-defined types (analogs of C++ classes) are named as variables (camel-case nouns), except that they begin with a capital letter:

```
Vec2D, OceanBox
```

Naming Rules for Procedures (Functions, Subroutines)

- Normal procedures (i.e., those which are not bound to a specific user-defined type) look similar to usual variables, *except* that they contain verbs, to emphasize the function of the procedure:

```
swap, isPrime, computeAverageTemperature
```

- Procedures that are bound to a specific type follow the rules above, but also have the name of the type at the end:

```
swapReal, getMagnitudeVec2D
```

This rule is introduced mostly to avoid naming collisions, when the same type-bound procedure name makes sense for several types (but their implementation differs). For simplifying the calling of these procedures, we usually define shorter aliases (which omit the type-name), as explained in Chap. 3.

Naming Rules for Modules and Source Code Files

- For naming Fortran modules which *do not* encapsulate a user-defined type, we use nouns and camel-case (first letter being uppercase):

```
Utilities, NumericKinds
```

A common guideline is to place each module in a separate file; for example, the modules above would be placed in files Utilities .f90 and NumericKinds.f90 , respectively. However, we do not adhere to this rule until later in the book, after explaining how to work with projects composed of multiple files, in Sect. 5.1.

- Fortran modules that also encapsulate a user-defined type are named after the type, with the added prefix $\boxed{\_\text{class}}$:

  $\boxed{\texttt{Vec2D\_class, OceanBox\_class}}$

  When these are placed in distinct files, the filename is composed of the module name, with the added extension $\boxed{.\,\text{f90}}$. For example, the modules above would be placed in files $\boxed{\text{Vec2D\_class.f90}}$ and $\boxed{\text{OceanBox\_class.f90}}$.

Chapter 1
General Concepts

This chapter introduces the Fortran programming language in the context of numerical modeling, and in relation to other languages that the reader may have experience with. Also, we discuss some technical requirements for making the best use of this book, and provide a brief overview of the typical workflow for writing programs in Fortran.

1.1 History and Evolution of the Language

In the 1950s, a team from *International Business Machines Inc.* (IBM) labs led by John Backus created the Fortran ("mathematical FORmula TRANslation system")[1] language. This was the first *high-level language* (HLL) to become popular, especially in the domain of numerical and scientific computing, for which it was primarily designed. Prior to this development, most computer systems were programmed in assembly languages, which only add a thin wrapper on top of raw machine language (generally leading to software which is not portable and more difficult to maintain). Fortran was widely adopted due to its increased level of abstraction, which made Fortran programs orders of magnitude more compact than corresponding assembly programs. This popularity, combined with intentional simplifications of the language (for example, lack of pointer type in earlier versions), encouraged the development of excellent optimizing compilers, making Fortran the language of choice for many demanding scientific applications.

This is also the case for *Earth system science* (ESS), where Fortran is to date the most used programming language. The reasons are simple: there is a huge body of tested Fortran routines, and the language is very suitable for coding physical

---

[1] The reader may sometimes encounter the name of the language spelled in all capitalized (as in FORTRAN), usually referring to the early versions of the language, which officially supported only uppercase letters to be used in programs. This shortcoming was corrected by the later revisions, with which we are concerned in the present text.

© Springer-Verlag Berlin Heidelberg 2015
D.B. Chirila and G. Lohmann, *Introduction to Modern Fortran*
for the Earth System Sciences, DOI 10.1007/978-3-642-37009-0_1

equations. Early model implementations based on Fortran started in the mid of last century (see e.g. Bryan [1], Platzman [4], Lynch [3] and references therein). The models predicted how changes in the natural factors that control climate, such as ocean and atmospheric currents and temperature, could lead to climate change. Climate models are intended to provide a user-friendly and powerful framework for simulating real or idealized flows over wide-ranging scales and boundary conditions. With its good support for modular programming, Fortran proved to be well suited for these tasks.

Certainly, many other languages were introduced over the last 60 years (such as the COBOL, Pascal, C, C++, Java, etc.), some offering innovative facilities for expressing algorithm abstractions (such as object-oriented or generic programming). Interestingly, these languages did not supersede Fortran (at least not in the ESS community); instead, they inspired the Fortran language-standardization committee to incorporate such facilities through incremental revisions (Fortran 90, Fortran 95, Fortran 2003, and Fortran 2008 at the time of writing).

1.2 Essential Toolkit (Compilers)

Fortran is a compiled language, so an ASCII text editor and a compiler should be enough to get started. A popular compiler is the *GNU Fortran Compiler* (gfortran), which is freely available as part of the *GNU Compiler Collection* (gcc). For users of Unix-like systems, this should be easily available, either in the system's package manager (*GNU/Linux* (Linux)), or bundled within the XCode developer package for *Mac OS X*® (OSX). It is also possible to install gfortran on *Microsoft Windows*® (Windows) systems, using the *Minimalist GNU for Windows* (MinGW) or Cygwin systems.

Many other compilers exist, some offering useful features like more powerful code optimizers,[2] convenient debugging/profiling tools, and/or a user-friendly *integrated development environment* (IDE). It is not possible to cover the whole landscape here—please consult a local expert or system administrator for advice on a suitable compiler.

The example programs were tested with recent versions of gfortran and of the *Intel Fortran Compiler*® (ifort), on Linux. However, the programs should be easy to adapt to other recent compilers and/or platforms.

---

[2] For most supercomputers, the compilers are usually provided by the hardware vendor, which allows better tuning of the code to the features of the underlying machine.

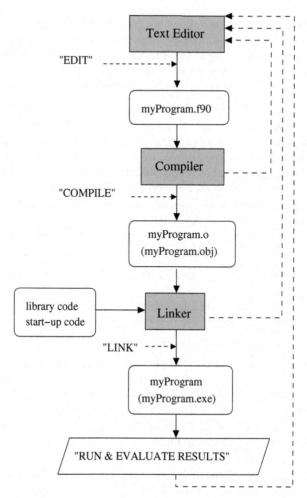

Fig. 1.1 Schematic of programming workflow in Fortran. Files are represented as *white rounded boxes*, and external programs as *green boxes*

1.3 Basic Programming Workflow

From a low-level perspective (i.e. leaving more abstract issues such as program design aside), development of Fortran programs[3] is represented schematically in Fig. 1.1.

---

[3] The terms "program" and "(source) code" are used interchangeably within this book; however, strictly speaking, "code" can also refer to program sub-modules, such as functions, while "program" usually refers to a complete application, which yields an executable file when processed by a compiler.

In the figure, the utilities are shown as green boxes. The process starts with a text editor,[4] where the user enters the program code.[5] Then, the compiler is invoked, passing the created file as an argument. In `Linux`, using `gfortran`, this would be achieved by typing the following command in a terminal window:

```
$ gfortran -c myProgram.f90
```

At this point, an additional file (`myProgram.o`) will be created. This contains machine code generated from `myProgram.f90` which does not contain, however, any code for libraries that may be needed by your program. It is the job of the linker to find the missing pieces and to produce the final, executable file. In `Linux`, the *GNU linker* (`ld`) is normally used for this purpose. For simplicity, it is better to perform the linking stage also through the compiler, which will call the linker with the appropriate options in the background:

```
$ gfortran -o myProgram myProgram.o
```

(in `Windows`, replace `myProgram.o` with `myProgram.obj`, and `myProgram` with `myProgram.exe`).

This step will create the executable program, which can be run with the command:

```
$ ./myProgram
```

The entire workflow seems deceivingly simple.[6] In reality, problems can appear at any stage (especially in nontrivial programs), which trigger the need to revise the program. These iterative improvements of the code are suggested by the dashed lines in Fig. 1.1. First, the compiler may refuse to produce object-code if the program does not follow the syntax of the language. Then, the linker may be unable to find the appropriate libraries to include. Finally, the program may crash, or it may run but produce unacceptable results. The beginner will usually encounter problems across all of these ranges. Fortunately, with some practice, the frequency of the (less interesting) compilation/linking errors decreases.

Compiling and linking in one step. So far, we separated the two phases for producing the program executable, to make the reader aware of the distinction (when

---

[4] Word processors are a poor choice here, since they focus on features like advanced formatting, which the compiler does not understand anyway; instead, a "bare bones" text editor, but with programming-related features like syntax highlighting and auto-completion, is recommended, for example: *gedit text editor* (`gedit`) or *Kate text editor* (`Kate`) are good starting points; *Vim text editor* (`Vim`), *GNU Emacs text editor* (`Emacs`) or *Joe's Own Editor* (`joe`) are more advanced choices, that may pay off on the longer term.

[5] Files containing modern Fortran source code usually have the extension `.f90`, but the reader may also encounter extensions `.f77`, `.f`, or `.for`, which correspond to older standards; likewise, some developers may use the extensions `.f95`, `.f03`, or `.f08`, to highlight use of features present in the latest revisions of the language—but this practice is discouraged by some authors (e.g. Lionel [2]). To avoid problems, filenames should also not contain whitespace.

[6] Indeed, this resembles the Feynman problem-solving algorithm: (a) write down the problem, (b) think very hard, and (c) write down the answer.

the executable fails to build properly, it is useful as a first step to determine if we face a compiler or linker error). However, for single-file programs, these steps can be combined in a single command:

```
$ gfortran -o myProgram myProgram.f90
```

For programs consisting of several files, compiling and linking by hand is impractical, and a build system becomes essential (discussed later, in Sect. 5.1).

References

1. Bryan, K.: A numerical method for the study of the circulation of the world ocean. J. Comput. Phys. **4**(3), 347–376 (1969)
2. Lionel, S.: Doctor Fortran in "Source Form Just Wants to be Free" (2013). http://software.intel.com/en-us/blogs/2013/01/11/doctor-fortran-in-source-form-just-wants-to-be-free
3. Lynch, P.: The origins of computer weather prediction and climate modeling. J. Comput. Phys. **227**(7), 3431–3444 (2008)
4. Platzman, G.W.: The ENIAC computations of 1950—gateway to numerical weather prediction. Bull. Am. Meteorol. Soc. **60**(4), 302–312 (1979)

Chapter 2
Fortran Basics

In this chapter, we introduce the basic elements of programming using Fortran. After briefly discussing the overall syntax of the language, we address fundamental issues like defining variables (of intrinsic type). Next we introduce *input/output* (I/O), which provides the primary mechanism for interacting with programs. Afterwards, we describe some of the flow-control constructs supported by modern Fortran (if, case, and do), which are fundamental to most algorithms. We continue with an introduction to the Fortran array-language, which is one of the strongest points of Fortran, of particular significance to scientists and engineers. Finally, the chapter closes with examples of some intrinsic-functions that are often used (for timing programs and generating pseudo-random sequences of numbers).

2.1 Program Layout

Every programming language imposes some precise syntax rules, and Fortran is no exception. These rules are formally grouped in what is denoted as a "context-free grammar",[1] which precisely defines what represents a valid program. This helps the compiler to unambiguously interpret the programmer's source code,[2] and to detect sections of source code which do not follow the rules of the language. For readability, we will illustrate some of these rules through code examples instead of the formal notation.

Below, we show the basic layout of a single-file Fortran program, with no procedures (these will be discussed later):

---

[1] For example, *extended Backus-Naur form* (EBNF).

[2] EBNF is also useful for defining consistent data formats and even simple *domain-specific languages* (DSLs).

© Springer-Verlag Berlin Heidelberg 2015
D.B. Chirila and G. Lohmann, *Introduction to Modern Fortran
for the Earth System Sciences*, DOI 10.1007/978-3-642-37009-0_2

```
program [program name]
  implicit none

  [ variable declarations [ initializations ] ]

  [ code for the program ]

end program [program name]
```

Any respectable language tutorial needs the classical "Hello World" example. Here is the Fortran equivalent:

```
program hello_world
  implicit none
  print*,"Hello, world of Modern Fortran!"
end program hello_world
```

Listing 2.1   src/Chapter2/hello_world.f90

This should be self-explanatory, except maybe for the implicit none entry, which instructs the compiler to ensure all used variables are of an explicitly defined type. It is strongly recommended to include this statement at the beginning of each program.[3] The same advice will apply to modules and procedures (discussed later).

> **Exercise 1** (*Testing your setup*) Use the instructions from Sect. 1.3 (adapting commands and compiler flags as necessary for your system) to edit, compile and execute the program above. Try separate compilation and linking first, then combine the two stages.

2.2 Keywords, Identifiers and Code Formatting

All Fortran programs consist of several types of *tokens*: *keywords* (reserved words of the language), *special characters*,[4] *identifiers* and *constant literals* (i.e. numbers, characters, or character strings). We will encounter some of the keywords soon, as we discuss basic program constructs. Identifiers are the names we assign to variables or constants. The first character of an identifier should be a letter (the rest can be

---

[3] This is related to a legacy feature, which could lead to insidious bugs. The take-home message for new programmers is to always use implicit none . The −fimplicit−none flag can be used, in principle, in gfortran, but this is also discouraged because it introduces an unnecessary dependency on compiler behavior.

[4] The special characters are (framed by boxes): =, +, -, *, /, (,), ,, ., $, ', :, ⬚ (blank), !, ", %, &, ;, <, >, ?, \, [,], {, }, ~, `, ^, |, #, and @. Certain combinations of these are reserved for *operators* and *separators*.

letters, digits or underscores $\boxed{\_}$). The length of the identifiers should not exceed 63 characters (Fortran 2003 or newer).[5]

Comments: Commenting the nontrivial aspects of your code is highly recommended.[6] In Fortran, this is achieved by typing an exclamation mark ($\boxed{!}$), either on a line of its own, or within another line which also contains program code. In either case, an $\boxed{!}$ will cause the compiler/preprocessor to ignore the rest of the line.[7]

Multi-line statements: Unlike languages from the C-family, in Fortran the semicolon $\boxed{;}$ for marking the end of a statement is optional (although it is still used sometimes, to pack several short statements on the same line). By default, the end of the line is also considered to be the end of the statement. A line of code in Fortran should be at most 132 characters long. If a statement is so long that this is not sufficient (for example, a long formula for evaluating derivatives in finite-difference numerical schemes), we can choose to continue it on the following line(s), by inserting an ampersand $\boxed{\&}$ at the end of each line that is continued. Since Fortran 2003, up to 255^8 continuation lines are allowed for any statement.

It can happen (although it should be avoided when possible) that the line break in a multi-line statement occurs at the middle of a token. In that case, using a single $\boxed{\&}$ will probably not give the expected result. This can be overcome by typing another $\boxed{\&}$ as the first character on the continued line, which contains the remainder of the divided token.

The two possible uses of continuation lines are shown in the example below:

```fortran
program continuation_lines
   implicit none
   integer :: seconds_in_a_day = 0

   ! Normal continuation-lines
   seconds_in_a_day = &
        24*60*60 ! 86400

   print*, seconds_in_a_day

   ! Continuation-lines with a split integer-literal token
   seconds_in_a_day = 2&
        &4*60*60 ! still 86400. In this case, splitting the '24'
                 ! is unwise, because it makes code unreadable.
                 ! However, for long character strings this can be
                 ! useful (see below).
   print*, seconds_in_a_day

   ! Continuation-lines with a split string token.
   print*,"This is a really long string, that normal &
        &lly would not fit on a single line."
end program
```

Listing 2.2 $\boxed{\text{src/Chapter2/continuation\_lines.f90}}$

---

[5] A maximum of 31 characters were allowed in Fortran 95.

[6] A good guideline is to make the code indicate clearly **what** is being done (through choice of meaningful variable and function names), and then to use the comments to describe the motivations (**why** it has been done like that, and what other problem-specific aspects are relevant).

[7] Exceptions to this rule are compiler directives ("pragmas"), which are specially-formatted comments that communicate additional information to the compiler; examples will be shown in Sect. 5.3, when we will discuss how to specify, using the *Open MultiProcessing* (OpenMP) extensions, which portions of the code should be attempted to be run in parallel.

[8] The previous limit (according to the Fortran 95 standard) was of up to 39 continuation lines.

Spaces and indentation: Whitespace can be freely used to separate program tokens, without changing the meaning of the program. For example, as far as the compiler is concerned, *line 3* in the previous listing could also have been written as:

```
integer              ::seconds_in_a_day=        0
```

Therefore, this is a subjective choice, which can be used to our advantage, to improve readability of the code. For example, it is considered good practice to indent[9] program-flow constructs (loops, conditionals, etc.), as will be shown later.

Combining statements in one line: As previously mentioned, we normally have one statement per line of code. However, it is also allowed to combine instructions, as long as they are separated by semicolons $\boxed{;}$. A common example is for swapping two variables ($\boxed{a}$ and $\boxed{b}$) using a temporary ($\boxed{\texttt{temp}}$):

```
temp=a; a=b; b=temp ! semicolon not mandatory at end of line
```


2.3 Scalar Values and Constants

As other programming languages, Fortran allows us to define named entities, for representing quantities of interest from the problem domain (speed, temperature, concentration of a tracer, etc.). Each entity belongs to a type, which specifies a *set of possible values*, a *scheme for encoding those values*, and *allowed operations*. Fortran is *statically-typed*, which means the type of a variable is fixed at compile-time.[10] This apparent drawback actually helps in practice, because many errors can be caught earlier; it also helps the compiler to apply certain code-optimization techniques (because the number of bits needed for each variable is known well before any operation is applied to the variable).

Standard-compliant compilers should provide at least five built-in types.[11] Of these, three are *numeric* ($\boxed{\texttt{integer}}$, $\boxed{\texttt{real}}$ and $\boxed{\texttt{complex}}$), and two *non-numeric* ($\boxed{\texttt{character}}$ and $\boxed{\texttt{logical}}$).

---

[9] Note that some text editors feature automatic indentation, which makes this easier.

[10] Other languages, such as *Matrix Laboratory*® (MATLAB) or *The R Project for Statistical Computing* (R), support *dynamic typing*, so the type of a variable can change during the execution of the program.

[11] It is also possible to define custom types, enabling data-encapsulation techniques similar to C++ (this will be discussed in Sect. 3.3.2).

All of these types can be used to declare *named constants* or *variables*:

```
! Declaring normal variables
! -- numeric --
integer :: length = 10
real    :: x = 3.14
complex :: z = (-1.0, 3.2)
! -- non-numeric --
character :: keyPressed ='a'
logical   :: condition = .false. ! (either'.true.' OR'.false.')

! Declaring named constants
! -- numeric --
integer, parameter :: INT_CONST = 30
real, parameter    :: REAL_CONST = 1.E2 ! (scientific notation)
complex, parameter :: I = (0.0, 1.0)
! -- non-numeric --
character, parameter :: B_CHAR ='b'
logical, parameter   :: IS_TRUE = .true.
```

NOTES

- **Position of declarations in (sub)programs:** All declarations for constants and variables need to be included at the beginning of the (sub)program, *before* any executable statement. However, as of Fortran 2008 it is possible to overcome this limitation, by surrounding variable declarations with a block — end block construct, as follows:

```
! variable declaration(s)
integer :: length

! executable statements (normally, not possible to specify additional
! variable declarations after the first such executable statement)
length = 10

block
    ! block-construct (Fortran 2008+) enables us to overcome that
    ! limitation
    real :: x
end block
```

- **The : : separator:** In the examples below, we declare variables both with and without this separator. In general, : : is optional, except when *the variable is also initialized* or *variable attributes are specified*. A simple rule of thumb is to always use this separator, which works in all cases.
- **Constants:** Any value can be declared as a constant, by appending the parameter -attribute after the name of the type. *This should be used generously whenever values are known to be constant, to avoid accidental overwriting.* Other type attributes will also be discussed.

2.3.1 Declarations for Scalars of Numeric Types

Below, we present some examples of defining variables and constants of numeric types. For each type, we first demonstrate how to define a variable. A definition only reserves space for the variable in memory, but the value stored in its corresponding bits is actually undefined, so it is highly recommended to follow each declaration with an initialization. These two steps can be merged into a one-liner (see examples below). Finally, we also show how to define constants of each type.

$\boxed{\texttt{integer type:}}$ valid values of this type are, e.g. -42, 24, 123. In general, any integer is accepted, as long as it resides within a certain range. The length of the range is determined by the `kind` parameter (if that is explicitly specified), or by the machine architecture (32 or 64 bit) and compiler (if no `kind` is specified, as in our present examples). Example declarations:

```
integer i              ! plain declaration...
i = 10                 ! ...with corresponding initialization
                       ! (would be in the executable section of
                       ! the (sub)program)
integer             :: j = 20 ! declaration with initialization
integer, parameter  :: K = 30 ! constant (initialization mandatory)
```

Note that, unlike other programming languages, Fortran `integer`-variables are always signed (i.e. they can be both positive and negative).

$\boxed{\texttt{real type:}}$ valid values of this type are, e.g. 2.78, 99., 1.27e2 (exponential notation)[12] or .123. Similar to integers, the number of digits after the decimal point (precision) and range of the exponent are system-and `kind`-dependent. Example declarations:

```
real x                 ! simple declaration
real             :: y = 1.23 ! declaration with initialization
real, parameter  :: z = 1.e2 ! constant (scientific notation)
```

$\boxed{\texttt{complex type:}}$ complex numbers are often needed in scientific and engineering applications, thus Fortran supports them natively. They can be specified as a pair of `integer` or `real` values (however, even if both components are specified as integers, they will be stored as a pair of reals, of default `kind`). Example declarations:

```
complex c1 ! simple declaration
complex :: c2 = (1.0, 7.19e23) ! declaration with initialization
complex, parameter :: C3 = (2., 3.37) ! constant
```

2.3.2 Representation of Numbers and Limitations of Computer Arithmetic

While internally all data is stored by computers as a sequence of bits (zeroes and ones), the concept of types (also known as "data types") binds the byte sequences to specific interpretations and manipulation rules. For example, addition of $\boxed{\texttt{integer}}$-versus that of $\boxed{\texttt{real}}$-numbers is very different at the bit-level. The number of bits used for a value of each type is particularly important: the more bits are used, the

---

[12] $1.27e2 \equiv 1.27 \times 10^2$.

more numbers become representable. For ⎡integer⎤s, this is exploited to increase
the bounds of the representable interval; for ⎡real⎤s, part of the extra bits can be
used to increase the *precision* (i.e. roughly the number of digits after the decimal
point, after we translate the number back to the decimal representation). As a rule of
thumb, computations become more expensive when more bits are used.[13] However,
numerical algorithms also vary with respect to the precision they need to function
correctly. To balance these factors, most computer systems support several sub-types
for ⎡integer⎤ and ⎡real⎤ values.

Modern Fortran has a very convenient mechanism for specifying the numerical
requirements of a program in a portable way, without forcing developers (or, worse,
users) to study each CPU in-depth. We discuss this feature in Sect. 2.3.4.

It is important that programmers keep in mind the limitations of the internal
representations, since these are an endless source of bugs. A tutorial on these issues
is outside the scope of our text (a very readable introduction to these issues and their
implications is Overton [11]). For example, some of the facts to keep in mind for the
⎡integer⎤ and ⎡real⎤ types are:

- ⎡integer⎤: Unlike C, Fortran always stores integer-values with a sign.[14]

  All ⎡integer⎤ types represent *exactly* all numbers within a sub-interval of the
  mathematical set of integer numbers. However, different kinds of integers (using
  different number of bits) will have different lengths for the representable interval.
  This is important when our programs use conversions from one kind of integer
  to another. Also, operations involving two integers may produce a result which
  is not representable inside the type (a situation known as *integer overflow*). Sometimes, compilers may have options which can detect such errors when a program
  is tested – for example, gfortran can achieve this when the ⎡−ftrapv⎤ flag
  is used.[15]

- ⎡real⎤: Most computer systems nowadays support the ⎡IEEE 754⎤ standard. This
  specifies a set of rules for representing fractional numbers inside the computer,
  along with bounds on the errors they introduce. This representation is also known
  as "floating-point", since it was inspired by the floating-point representation of
  large numbers, used in science and engineering calculations. While integer-
  arithmetic is exact (as long as both the arguments and the result are representable),
  this is not the case for floating-point representations: since any interval along
  the real axis contains an infinite set of numbers, it is impossible to store most

---

[13] This is not a "hard" rule, however, because many factors enter the performance equation—
e.g. specialized hardware units within the *central processing unit* (CPU), the memory hierarchy,
vectorization, etc.

[14] An intuitive approach would be to reserve one bit for the sign, and use the rest for the modulus.
However, to reduce hardware complexity most systems use another convention ("two's complement").

[15] Note that enabling such options will most probably make the program slower too, so they are not
meant for "production" runs.

numbers using a bit-field of finite size. This causes most nontrivial calculations with `real`-values in Fortran to be *approximate* (so there is always some "noise" in the numerical results).

To complicate matters more, note that *many numbers which are exactly representable in the familiar decimal floating-point notation* cannot *be represented exactly when translated to one of the binary floating-point formats*. A common example is the number 0.1, which on our system becomes 0.100000001490116 when translated to 32 *bit* floating-point and back to decimal, and 0.10000000 0000000005551115123125783 when 64 *bit* floating-point is used. This can lead to subtle bugs—for example, when two variables which were both assigned the value 0.1 are compared for equality, the result may be *false* if the two variables are of different floating-point type. In this case, the compiler will promote the lower-precision value to the higher-precision type (so that it can perform the comparison). However, this cannot bring back the bits that were discarded when 0.1 was converted to fit inside the lower-precision type. For this reason, it is often a good idea to avoid such comparisons as long as possible, or to include some tolerances when this operation is necessary nonetheless. For more information on floating-point arithmetic and advice for good practices, the reader can also consult Goldberg [4], as well as Overton [11].

2.3.3 *Working with Scalars of Numeric Types*

The three numeric types share some characteristics, so it makes sense to discuss their usage simultaneously, highlighting any exceptions. This is the purpose of this section.

2.3.3.1 Constructing Expressions

Scalars of numeric type (*operands*), together with *operators*, can be combined to form *scalar expressions*. Fortran supports the usual *binary* arithmetic operators: $\boxed{**}$ (exponentiation), $\boxed{*}$ (multiplication), $\boxed{/}$ (division), $\boxed{+}$ (addition), and $\boxed{-}$ (subtraction). The last two may also be used as *unary* operators (for example, to negate a value). Complex expressions can be built, with more than one operator. For evaluation, these are divided into sub-expressions, which are executed in left-to-right order, with some complications due to the precedence rules (for the details, consult, for example, Metcalf et al. [10]). Parentheses can be used to override the precedence rules, which may make code readable in some cases:

```
real :: x=13, y=17, z=0
z = x*y+x/y     ! this expression (using precedence rules)
z = (x*y) + (x/y) ! is equivalent to this one (using parentheses)
```

2.3.3.2 Mixed-Mode Expressions

The generality of numeric types in Fortran mirrors their mathematical counterparts: $\mathbb{Z} \subset \mathbb{R} \subset \mathbb{C}$. When operands in a numeric expression do not have the same type and kind, Fortran usually converts the less precise/general operand, to conform to the more precise/general of the operands. A notable exception to this rule is when raising a real to an integer-power, in which case the integer is not converted to real (which is good, since raising to an integer power is more accurate and faster than raising to a corresponding real power). Another, less fortunate exception is when one of the operands is a literal constant, which can lead to loss of precision (therefore it is recommended to ensure the kind of the constant is specified—how to do this will be shown in Sect. 2.3.4).

2.3.3.3 Using Scalar Expressions

Standalone numerical expressions do not make much sense (hence the language does not allow them): what we actually want is to *assign* the result of the expressions to variables (with the $\boxed{=}$ assignment operator[16]), or to *pass* the result to some function (e.g. to display it). This is another point where loss of precision can occur, if the expression is of a stronger type/kind than the variable to which it is assigned:

```
integer :: i = 0
real :: m = 3.14, n = 2.0
i = m / n    ! i will become 1, NOT 1.57 (rounding towards 0)
m = -m       ! negate m with unary operator
i = m / n    ! i will become -1 (rounding also towards 0)
print*, m / n ! expression passed to 'print'-function
```

2.3.3.4 Convenient Notation for Sub-components of complex

For applications that need to work with data of complex type, note that it is possible (since Fortran 2008) to conveniently refer to the real and imaginary components:

```
complex :: z1(1.0, 2.0)
z1%im = 3.0 ! modify the imaginary part
print*,"real part of z1 = ", z1%re
```

2.3.4 The kind type-parameter

Most of the numerical algorithms encountered in ESS need some assumptions regarding properties of the types used to represent the quantities they manipulate. For example, if integers are used to represent simulation time in seconds, we need to ensure the type can support the maximum number of seconds the model will be run for. The

---

[16] Not to be confused with an equivalence in the mathematical sense. In Fortran, that is represented by the $\boxed{==}$ operator, which we discuss shortly, in relation to the logical type.

demands are more complex for reals, which are always stored with finite precision:
since each result needs to be truncated to fit the representation, numerical "noise" is
ever-present, and needs to be constrained.

One way to improve[17] the situation is to increase the accuracy of the representation, by using more bits to represent each value. In older versions of Fortran, the
`double precision` type (`real`-variant) was introduced exactly for this. The
problem, however, lies in the fact that the actual number of bits is still system- and
compiler-dependent: switching hardware vendors and compilers is normal, and surprises due to improper number representation (which can often go unnoticed) are
better avoided when possible.

The concept of `kind` is the modern Fortran response to this problem, and it
deprecates the `double precision` type. `kind` acts as a parameter to the type,
allowing the programmer to select a specific type variant from the multitude that may
be supported by the platform.[18] Even better, the programmer need not be concerned
with the lower-level details, since two special intrinsic functions (discussed shortly)
allow querying for the most economic types that meet some natural requirements.

We only discuss `kind` for numeric types, although the concept also applies to non-numeric types (for storing characters of non-European languages and for efficiently
packing arrays of `logical` type—for details consult, e.g. Metcalf et al. [10]).

Kinds are indexed with positive integer values. If we know these indices for
selecting numbers with the desired properties on the current platform, they can be
used to parameterize types, as in:

```
integer( kind=4) :: i
real( kind=16) :: x
```

However, this feature alone does not solve the portability problem, since the index
values themselves are not standardized. The intended usage, instead, is through two
intrinsic functions which return the appropriate `kind`-values, given some constraints
requested by the developer:

1. `selected_int_kind` (`requestedExponentRange`), where `request-
   edExponentRange` is an integer, returns the appropriate `kind`-parameter for
   representing integer numbers in the range:

$$-10^{requestedExponentRange} < number < 10^{requestedExponentRange}$$

For example,[19]

```
integer, parameter :: LARGE_INT = selected_int_kind(18)
integer(kind=LARGE_INT) :: t = -123456_LARGE_INT
```

---

[17] This is not to be seen as a "silver bullet", since numerical noise will still corrupt the results, if
the algorithm is inherently unstable.

[18] Compilers are required to provide a default `kind` for each of the 5 intrinsic types, but they may
(and most do) support additional kinds.

[19] In practice, it is more convenient to use shorter denominators for the `kind`-parameters.

will guarantee that the compiler selects a suitable type of `integer` to fit values of t in the interval $(-10^{18}, 10^{18})$.

2. `selected_real_kind(requestedPrecision, requestedExpo-nentRange)`, where both arguments are integers, returns the appropriate `kind`-parameter for representing numbers with a *decimal exponent range* of at least `requestedExponentRange`, and a *decimal precision* of at least `reques-tedPrecision`[20] after the decimal point.

Example:

```
integer, parameter :: MY_REAL = selected_real_kind(18,200)
real(kind=MY_REAL) :: x = 1.7_MY_REAL
```

To obtain what is commonly denoted as single-, double-, and quadruple-precision, the following parameters can be used:

```
integer, parameter :: R_SP = selected_real_kind(6,37)
integer, parameter :: R_DP = selected_real_kind(15,307)
integer, parameter :: R_QP = selected_real_kind(33,4931)
```

Note that, since the exact data type needs to be revealed to the compiler, results of the `kind`-inquiries need to be stored into *constants* (which are initialized at compile-time).

By increasing the values of the `requestedExponentRange` and/or `requestedPrecision` parameters, it is easily possible to ask for numbers beyond the limits of the platform (you will get the chance to test this in Exercise 7). In such situations, the inquiry functions will return a negative number. This fits with the way `kind` type-parameters are used, since trying to specify a negative `kind` value will cause the compilation to fail:

```
integer, parameter :: NONSENSE_KIND = -1
integer(kind=NONSENSE_KIND) :: s ! will fail to compile

integer, parameter :: UNREASONABLE_RANGE = selected_int_kind(30000)
! will also fail to compile (at least, in 2013), because a too ambitious range
! of values was requested, causing the intrinsic function to
! return a negative number
integer(kind=UNREASONABLE_RANGE) :: u
```

In closing of our discussion on `kind`, we have to admit that inferring the type-parameters in each (sub)program, while viable for simple examples, can become tedious and, worse, leads to much duplication of code. An elegant solution to this problem is to package this logic inside a `module`, which is then included in (sub)programs.[21] We defer the discussion of this mechanism to Sect. 3.2.7, after covering the concept of modules.

---

[20] The situation is more complex for this type, because some values which are exactly representable using the *decimal* floating-point notation can only be approximated in the *binary* floating-point notation.

[21] We encountered this mechanism in the Fortran distribution of the popular *Numerical Recipes* book, see Press et al. [12].

2.3.5 Some Numeric Intrinsic Functions

As a language designed for science and engineering applications, Fortran supports a large suite of mathematical functions, which complement the operators. Also, including these as part of the core language allows vendor-specific implementations to take advantage of special hardware support for some costly functions.

Among the most frequently-used numeric intrinsic functions, we mention:

- *type conversion*: `real(x [, kind])`, `int(x [, kind])`
- *trigonometric functions* (operating with radians): `sin(x)`, `cos(x)`, `tan(x)`; also, inverse (`asin(x)`, `acos(x)`, `atan(x)`), and hyperbolic functions (`sinh(x)`, `cosh(x)`, `tanh(x)`)
- *usual mathematics*: `abs(x)` (absolute value), `exp(x)`, `log(x)` (natural logarithm), `sqrt(x)`, `mod(x [,n])` (remainder modulo-n)

This list is by no means comprehensive (see Metcalf et al. [10] for an exhaustive version). The `kind` of the result is usually the same as that of the first parameter, unless the function accepts a `kind`-parameter, which is present.

In Sect. 2.7, we discuss some more intrinsic functions, useful for more advanced tasks.

2.3.6 Scalars of Non-numeric Types

`logical type:` allows variables to take only two values: `.true.` or `.false.` (dots are mandatory). They can be declared similarly to the other types:

```
logical activated      ! plain declaration...
activated = .true.     ! ...with corresponding initialization
logical :: conditionSatisfied = .false. ! declaration with init
logical, parameter :: ON = .true. ! constant (init mandatory)
```

`logical` **expressions**: As for numeric types, `logical` values can be used, together with specific operators (unary: `.not.`; binary: `.and.`, `.or.`, `.eqv.` (equality) and `.neqv.`), to construct expressions, as in (using the previous declarations):

```
.not. conditionSatisfied     ! .eqv. .true.
conditionSatisfied .and. ON ! .eqv. .false.
```

It is important to know that `logical` expressions can also be constructed out of numeric arguments, using the arithmetic *comparison operators*: `==` (equal), `/=` (not equal), `>` (greater), `>=` (greater-or-equal), `<` (smaller), and `<=` (smaller-or-equal). Such `logical` expressions are used in flow-control statements (e.g. `if`), discussed in Sect. 2.5.

$\boxed{\texttt{character type:}}$ variables and constants of this type are used for storing text characters, similar to other programming languages. In Fortran, characters and character strings are marked by *a pair of single quotes* (as in '`some text`'), or *a pair of double quotes* (as in ''`some more text`''). These can be used interachangeably, both for single- and multi-character values.

A text character is said to belong to a character set. A very influential such character set is ASCII, which can be used to represent English-language text. Ours being an English text, we devote more space to this character set.

Many modern Fortran implementations currently use ASCII by default. For example, this is the case on our test system (64bit Linux, gfortran v4.8.2), when we declare variables such as:

```
character char1   ! plain declaration (to be initialized later)
character :: char2 ='a'  ! declaration with immediate initialization
```

We discussed earlier (in the context of numeric types) the concept of type-parameters. The character type actually accepts two such parameters: $\boxed{\texttt{len}}$ (for controlling the length of the string) and $\boxed{\texttt{kind}}$ (for selecting the character set).

Let us focus on the first parameter (len) for now. It exists because most of the times developers want to store *sequences of characters* (*strings*). If (as in the previous listing) len is not explicitly mentioned, it implicitly has the value fixed to "1" (reserving space for just one ASCII-character). To store wider strings, we can declare a sufficiently-large value for len, e.g.:

```
character(len=100) myName  ! fixed-size string
```

However, this method is not so convenient in practice.

For the case where the length of the string can be determined during compilation (i.e. it will not change when our program will be executed), we can use *assumed-length* strings. This is particularly useful for declaring constant strings, sparing the developer from counting characters (for the len parameter):

```
1  program assumed_length_strng_constant
2    implicit none
3
4    character(len=*), parameter :: FILENAME = & ! character constant
5      'really_long_file_name_for_which_&
6        &we_do_not_want_to_count_characters.out'
7
8    print*,'string length:', len(FILENAME)
9  end program assumed_length_strng_constant
```

Listing 2.3 $\boxed{\texttt{src/Chapter2/assumed\_length\_strng\_constant.f90}}$

Note the type-parameter $\boxed{\texttt{len=*}}$ (*line 4*), which causes the string to have what is known as *assumed-length*.

Another common scenario is when the strings to operate on are not constant, with their lengths only becoming known during the execution of the program. This is the case, for example, if we want to make the previous listing more flexible, by asking the user to provide a filename.[22] For such a situation, we can use *deferred-length* strings, which are marked by the type-parameter $\boxed{\texttt{len=:}}$, in conjunction with the specifier $\boxed{\texttt{allocatable}}$. For example:

```fortran
program deferred_length_strng
  implicit none

  character(len=256) :: buffer ! fixed-length buffer
  character(len=:), allocatable :: filename ! deferred-length string

  print*,'Please enter filename (less than 256 characters):'
  read*, buffer  ! place user-input into fixed buffer

  filename = & ! copy from buffer to dynamic-size string
      trim(adjustl(buffer))  !'trim' and'adjustl' explained later

  print*, filename  ! some feedback...
end program deferred_length_strng
```

Listing 2.4 $\boxed{\texttt{src/Chapter2/deferred\_length\_strng.f90}}$

It is not possible to place a value in `filename` directly from the `read`-statement (*line 8*). Therefore, we declare an extra buffer to hold the input data (*line 4*). The actual *deferred-length* variable is declared at *line 5*. On *line 7* we announce to the user that a string (i.e. characters surrounded by single or double quotes!) is expected, and on *line 8* we read the input into the buffer. At *line 10* we finally get to use the deferred-length variable. Ignoring the intrinsic functions for now, the net effect is that a string will be assigned to $\boxed{\texttt{filename}}$. Note that the system automatically reserves memory internally, so that our variable $\boxed{\texttt{filename}}$ is large enough. Later, we will also discuss how to *explicitly* request such memory, in the context of dynamic arrays (Sect. 2.6.8).

`character` **operators and intrinsic functions:** For the `character` type it is useful to know about the operator $\boxed{\texttt{//}}$, which concatenates two strings. Expressions formed with this operator are strings with length equal to the sum of the lengths of the strings to be concatenated. We usually want to assign the evaluated expressions to other string variables, in which case truncation or whitespace-padding (on the right) can occur, depending on the length of the expression and of the string variable we assign to. These situations are illustrated in the following example[23]:

```fortran
program character_type_examples
  implicit none

  ! given two source string-variables:
  character(len=4) :: firstName ="John"
  character(len=7) :: secondName ="Johnson"
  ! and 3 target variables (of different lengths):
  character(len=13) :: exactFit
  character(len=10) :: shorter
  character(len=40) :: wider =&
      "Some phrase to initialize this variable."
```

---

[22] This approach is more convenient, in the sense that the user does not have to re-compile the program every time the filename changes. For real-world software, we prefer to minimize interaction with users, and allow specification of filenames (e.g. model input) at the invocation command line instead (Sect. 5.5.1), which facilitates unattended runs.

[23] If you try this example, you may notice that an additional space is printed at the beginning of every line. This is the default behavior, related to some legacy output devices. We will discuss how to avoid this in Sect. 2.4.

```
! below, we concatenate'firstName' and'secondName',
! assigning the result to strings of different sizes.
! note:'/'-characters serve as markers, to highlight
! the spaces in the actual output.

! expression fits exactly into'exactFit'
exactFit = firstName //","// secondName
print*,"|", exactFit,"|"

! expression does not fit into'shorter', so some
! characters at the end are truncated
shorter = firstName //","// secondName
print*,"|", shorter,"|"

! expression takes less space than available in
!'wider', so whitespace is added as padding on
! the right (previous content discarded)
wider = firstName //","// secondName
print*,"|", wider,"|"
end program character_type_examples
```

Listing 2.5 | `src/Chapter2/character_type_examples.f90`

Table 2.1 Some intrinsic functions for `character`(-strings)

Function name	Result/Effect
`lge(string1, string2)`	`.true.` if `string1` follows after or is equal to `string2`
(similar: `lgt, lle, llt`)	(lexical comparison, based on ASCII collating sequence)
`len(string)`	length of `string`
`trim(string)`	`string`, excluding trailing padding-whitespace
`len_trim(string)`	length of `string`, excluding trailing padding-whitespace
`adjustr(string)`	right-justify `string`
(similar: `adjustl`)	

In ESS models, characters and strings are often secondary to the core numerics. They are, however, useful for manipulating model-related *metadata*. To cater for such needs, Fortran provides several intrinsic functions that take strings arguments (see Table 2.1 for a basic selection, or Metcalf et al. [10] for detailed information).

At the end of Sect. 2.5, after introducing more language elements, we use some of these intrinsic functions, to solve a common pattern in ESS (creation of unique filenames for transient-phenomena model output, based on the index of the time step).

2.4 Input/Output (I/O)

The I/O system is an essential part of any programming language, as it defines ways in which our programs can interact with other programs and with users.

For example, models in ESS typically read files (input) for setting-up the geometry of the problem and/or for loading initial conditions. Then, as the prognostic variables are calculated for the subsequent time step, the new model state is regularly written to other files (output). Frequently, the input files are created in an automatic fashion, using other programs; likewise, the output of the model may be passed to post-processing/visualization tools, for analysis.[24]

External files are not the only medium for performing I/O; other interfaces include the usual interaction with the user via the terminal, or communication with the *operating system* (OS) (which allows the program to become aware of command line arguments passed to it, and of environment variables—see Sect. 5.5 for some examples). It is also possible to construct *graphical user-interface* (GUI)-based I/O applications, using third-party libraries.[25] but, in ESS, models providing such features[26] are still the exceptions rather than the rule.[27]

We already used some simple I/O-constructs, in the code samples presented so far. In this section, we provide the background for these constructs, and also discuss other aspects of *formatted*[28] I/O (such as controlling the I/O commands, or working with files). Finally, we provide a hierarchical overview of the I/O facilities used in ESS.

NOTE

A distinguishing characteristic of Fortran is that, by default, its I/O subsystem is record-based *(unlike languages like C or C++, which treat I/O as a stream of bytes[a]).*

---

[a] This difference can cause problems while exchanging files across languages. Such problems can be avoided by using portable formats like *NETwork Common Data Format* (netCDF) (Sect. 5.2.2) or, when the file format cannot be changed, by using the new stream I/O capabilities of Fortran 2003 (see Metcalf et al. [10]).

2.4.1 List-Directed Formatted I/O to Screen/from Keyboard

The simplest form of I/O in Fortran, which we have used so far, enables communication with the program from the terminal where the program was launched. Here, data needs to be converted between the internal representation of the format, and the

---

[24] The complete network of tasks for obtaining the final data products can become quite complex. In such cases, it often pays off to automatize the entire process, using shell scripts (see Sect. 5.6.1 for a brief overview of the options available, and some suggestions for further reading).

[25] See, for example, *Java Application Programming Interface* (JAPI) for an open-source solution; a commercial alternative is Winteracter.

[26] A model which provides a GUI is the Planet Simulator (see Fraedrich et al. [3], Kirk et al. [7]).

[27] A lack of graphical interfaces does not imply obsolete software practices: textual, command line interfaces can be readily used to automate complete workflows. This paradigm is suitable for ESS models, which usually need a long time to run. However, GUI-based systems are often suitable for steering operations which complete very fast, such as low-resolution models or tools in exploratory data analysis.

[28] In Fortran, *formatted* I/O means ASCII-text form; conversely, *un-formatted* I/O means binary form. We do not cover binary I/O in this text, even if it is more space-efficient, due to possible portability issues (we highlight an alternative form of efficient I/O in Sect. 5.2.2).

character strings recognized by the terminal. The programmer would often want to control this conversion process, to achieve the desired formatting.[29] However, for testing purposes, the `read*` and `print*` forms can be used, known as *list-directed I/O*. These are demonstrated in the following program, which expects the user to enter a name and date of birth (year, month, day), and returns the corresponding day of the week:

```fortran
program birthday_day_of_week
    implicit none
    character(len=20) :: name
    integer :: birthDate(3), year, month, day, dayOfWeek
    integer, dimension(12) :: t = &
            [ 0, 3, 2, 5, 0, 3, 5, 1, 4, 6, 2, 4 ]

    print*,"Enter name (inside apostrophes/quotes):"
    read*, name
    print*,"Now, enter your birth date (year, month, day):"
    read*, birthDate

    year = birthDate(1); month = birthDate(2); day = birthDate(3)

    if( month < 3 ) then
        year = year - 1
    end if

    ! Formula of Tomohiko Sakamoto (1993)
    ! Interpretation of result: Sunday = 0, Monday = 1, ...
    dayOfWeek = &
            mod( (year + year/4 - year/100 + year/400 + t(month) + day), 7)

    print*, name," was born on a "
    select case(dayOfWeek)
    case(0)
        print*,"Sunday"
    case(1)
        print*,"Monday"
    case(2)
        print*,"Tuesday"
    case(3)
        print*,"Wednesday"
    case(4)
        print*,"Thursday"
    case(5)
        print*,"Friday"
    case(6)
        print*,"Saturday"
    end select

end program birthday_day_of_week
```

Listing 2.6 `src/Chapter2/birthday_day_of_week.f90`

The part of the I/O statements following the comma is called an *I/O list*. For input, this needs to consist of variables (also arrays), while for output any expression can be used.

Previously, we mentioned the record metaphor used by Fortran; this needs to be considered while feeding input at the terminal for a `read`-statement: each statement expects its input to span (at least one) distinct line (=record), so before any subsequent `read*`-statement is executed, the file "cursor" would be advanced to the next record, making it impossible to enter on a single line input for adjacent `read*`-statements.

---

[29] This is discussed later, in Sect. 2.4.2; the process is controlled via an *edit descriptor*, which is embedded in a *format specification*.

Thus, it is perfectly acceptable to write something like:

```fortran
program read_3variables_on_a_line
  implicit none
  character(len=100) :: station_name ! fixed-length, for brevity
  integer :: day_of_year
  real :: temperature

  read*, station_name, day_of_year, temperature
  ! provide feedback (echo input)
  print*,"station_name=", trim(adjustl(station_name)), &
      ", day_of_year=", day_of_year, &
      ", temperature=", temperature
end program read_3variables_on_a_line
```

Listing 2.7 | src/Chapter2/read_3variables_on_a_line.f90 |

providing as input:

```
'Bremerhaven/Germany' 125 8<Enter>
```

On the other hand, if the input is performed as in the following program:

```fortran
program read_3variables_on_3lines
  implicit none
  character(len=100) :: station_name ! fixed-length, for brevity
  integer :: day_of_year
  real :: temperature

  read*, station_name
  read*, day_of_year
  read*, temperature
  ! provide feedback (echo input)
  print*,"station_name=", trim(adjustl(station_name)), &
      ", day_of_year=", day_of_year, &
      ", temperature=", temperature
end program read_3variables_on_3lines
```

Listing 2.8 | src/Chapter2/read_3variables_on_3lines.f90 |

we need to split the data for the three variables over three lines (records), as in:

```
'Bremerhaven/Germany'<Enter>
125<Enter>
8<Enter>
```

As previously mentioned, this form of I/O is not recommended for anything but quick testing, because it is limited from two points of view:

1. *system-dependent format*: the system will ensure that all data is visible, but the outcome is frequently not satisfying, due to generous whitespace-padding, which may often decrease readability; we discuss how to resolve this issue in Sect. 2.4.2.
2. *fixed I/O-channels*: input is only accepted from the keyboard, and output will be re-directed to the screen.[30] This becomes counter-productive as soon as the volume of I/O increases; we discuss how to route the I/O-channels to files in Sect. 2.4.3.

---

[30] For C/C++ programmers: this is the Fortran equivalent to the stdin, stdout and stderr streams.

Exercise 2 (*Emission temperature of the Earth*) The simplest *energy balance model* (EBM) for computing the emission temperature (T_e) of the Earth (as observed from space) consists of simply equating the absorbed solar energy and the outgoing blackbody radiation (assumed isotropic). This gives (Marshall and Plumb [8]) the following equation:

$$T_e = \sqrt[4]{S_0 \frac{(1 - \alpha_p)}{4\sigma}}$$

where $\sigma = 5.67 \times 10^{-8}\,\mathrm{Wm}^{-2}\,\mathrm{K}^{-4}$ is the Stefan-Boltzmann constant and, for present-day, the average Earth albedo $\alpha = 0.3$ and the annualy-averaged flux of solar energy incident on the Earth is $S_0 = 1367\,\mathrm{Wm}^{-2}$.

Write a program which evaluates this equation, computing T_e. How does the result change if S_0 is 30 % lower? What about increasing α by 30 %?

2.4.2 Customizing Format-Specifications

Fortran allows precise control on how data is converted between the internal representation and the textual representation used for formatted I/O. This is achieved by specifying a *format specification*. In fact, the language provides three ways of specifying the format:

1. *asterisk* ($\boxed{*}$): this is what we have used so far. The effective format is platform-dependent.
2. *a character string expression* (*of default kind*): this consists of a list of *edit descriptors*, as will be explained in this section.
3. *a statement label*: this allows writing the format on a separate, labeled statement—a technique that may be useful for structuring I/O statements better. However, we do not emphasize this option, since the same effect can be obtained with character strings.

The basic form of the *output* statement is:

```
print <format> [, <I/O list>]
```

Similarly, the *input* statement looks like:

```
read <format> [, <I/O list>]
```

The *format*-part, on which we focus in this section, is usually a character expression of the form:

```
'( edit_descriptor_1, edit_descriptor_2, ... )'
OR
"( edit_descriptor_1, edit_descriptor_2, ... )"
```

where each edit descriptor in the comma-separated list corresponds to one or more[31] item(s) in the *I/O list* of the statement.

The task of the edit descriptor is to precisely specify how data is to be converted from the internal representation to the character representation external to the program (or the other way around). Fortran supports three types of edit descriptors, which can be combined freely: *data*, *character string*, and *control*.

Data edit descriptors: This is the most important category, since it refers to the actual data-conversion process. Such edit descriptors are composed of combinations of characters and positive integers, as discussed shortly. In general, the numbers represent the lengths of the different components in the text representation on the external device side. *For output of numeric types, a set of asterisks is printed if these lengths are too small for representing the value.*

Fortran provides different types of edit descriptors, for each of the intrinsic types.[32] We present them below, using monospace-font for characters that need to be typed literally, and *italic*-font for variables to be replaced by integer values. *Note that characters like* $\boxed{-}$ *(negation),* $\boxed{\cdot}$ *(decimal point) and* $\boxed{\text{e}}$ *or* $\boxed{\text{E}}$ *(marker for exponent), when they appear, are also accounted for in the values of the various field-width variables.*

- integer: either $\boxed{\texttt{i}w}$ or $\boxed{\texttt{i}w.m}$ may be used, where w specifies the width of the field, and m specifies that, on output, at least m digits should be printed *even if they are leading zeroes* (on input, the two forms are equivalent). Example:

```
integer :: id = 0, year=2012, month=9, day=1
integer, dimension(40) :: mask = 10
print*,"Enter ID (integer < 1000):"
read'(i3)', id
! echo id (with leading zeroes if < 100)
print'(i3)', id
! using multiple edit descriptors
print'(i4, i2.2, i2.2)', year, month, day
```

When the magnitude of the integers to be written is not known in advance, it is convenient to use the $\boxed{\texttt{i0}}$ edit descriptor, which automatically sets the field-width to the minimum value that can still contain the output value[33]:

---

[31] It is possible, and sometimes useful, to have less edit descriptors than elements in the I/O list. In such situations, the edit descriptors are reused, while switching to a new record (for examples, see Sect. 2.6.5).

[32] Special facilities also exist for arrays and derived types. We discuss the former in Sect. 2.6.5, after introducing the corresponding language elements. For the latter, see Metcalf et al. [10].

[33] This form is highly recommended, as it relieves the programmer from bugs associated with manually selecting the field width (corrupted, asterisks-filled fields can occur if the number of digits in the number exceeds the expected field width). However, this makes the formatting of values variable, and may not be appropriate for applications where precise control of alignment is important (like compatibility with other programs, or for improving the clarity of the output). Also, note that this approach *does not work* for *input* (where $\boxed{\texttt{i0}}$ would cause any input value to be set to zero).

```
print'(i0)', testInt ! works correctly for any value
```

Binary, octal, and hexadecimal (hex) integers: For some applications, it can be useful to read/write `integer`-values in a non-decimal numeral system (bases 2, 8, and 16 being the most frequent). This is easily achieved in Fortran, by replacing the i with b (for binary), o (for octal) and z (for hexadecimal) respectively. The field-width can also be specified or auto-determined, just like when using the decimal base. The following program uses such edit descriptors to convert decimal values to other bases (some new elements in the program will be covered later):

```
program int_dec_2_other_bases
  implicit none
  integer :: inputInteger

  ! elements of this will be clarified later
  write(*,'(a)', advance='no')"Enter an integer: "
  ! get number (field width needs to be manually-specified)
  read'(i20)', inputInteger
  ! (string in format discussed later) print...
  print'("binary: ", b0)', inputInteger !...min-width binary
  print'("octal : ", o0)', inputInteger !...min-width octal
  print'("hex   : ", z0)', inputInteger !...min-width hex
end program int_dec_2_other_bases
```

Listing 2.9 `src/Chapter2/int_dec_2_other_bases.f90`

- `real`: no less than seven types of edit descriptors are available for this type (reflecting Fortran's focus on numerical computing): $fw.d$, $ew.d$, $ew.dee$, $esw.d$, $esw.dee$, $enw.d$, $enw.dee$, where w denotes the total width of the field, d the number of digits to appear after the decimal point, and e (when present) the number of digits to appear in the exponent field.

The first type of edit descriptor (based on f) is appropriate when the domain of the values includes the origin, and does not span too many orders of magnitude (say, $0.01 \lesssim x \lesssim 1000$). Otherwise, the e-variants, which use exponential notation, are usually preferred. The different e-variants were introduced for supporting various conventions for representing floating-point values used in different fields. The distinction lies mainly in the way they scale the exponent, which correlates to the range of the significant (= the rest of the number, after excluding the exponent). This is summarized in Table 2.2 below.

Table 2.2 Prefixes for exponential notation in edit descriptors for `real`

Prefix	Resulting range for absolute value of significant
e	[0.1, 1.0)
en ("engineering")	[1, 1000)
es ("scientific")	[1, 10)

Similar to `integer`-values, w can be set to zero when performing output, causing a minimum field-width to be selected, which can still contain the significant.[34] However, this is not allowed for the $\boxed{e}$-variants.

The following program demonstrates the effects of different edit descriptors for writing `real`-values:

```fortran
program edit_descriptors_for_reals
  implicit none
  ! get kind for high-precision real
  integer, parameter :: QUAD_REAL = selected_real_kind(33,4931)
  real(kind=QUAD_REAL) :: testReal

  write(*,'(a)', advance='no')"Enter a real number: "
  read'(f100.50)', testReal
  ! print with various edit-descriptors
  print'(a, f0.2    )',"f0.2      :  ", testReal
  print'(a, f10.2   )',"f10.2     :  ", testReal
  print'(a, f14.4   )',"f14.4     :  ", testReal
  print'(a, e14.4   )',"e14.4     :  ", testReal
  print'(a, e14.6e3 )',"e14.6e3   :  ", testReal
  print'(a, en14.4  )',"en14.4    :  ", testReal
  print'(a, en14.6e3)',"en14.6e3  :  ", testReal
  print'(a, es14.4  )',"es14.4    :  ", testReal
  print'(a, es14.6e3)',"es14.6e3  :  ", testReal
end program edit_descriptors_for_reals
```

Listing 2.10 $\boxed{\text{src/Chapter2/edit\_descriptors\_for\_reals.f90}}$

- `complex`: can be formatted using pairs of edit descriptors for `real` values.
- `logical`: supports the $\boxed{lw}$ edit descriptor, where w denotes the width of the field (if $w = 1$, $\boxed{T}$ or $\boxed{F}$ are supported, while $w = 7$ enables support for the expanded notation of boolean values, i.e., $\boxed{.true.}$ and $\boxed{.false.}$). According to the language standard, the width is mandatory.
- `character` strings: can be used with the $\boxed{a}$ or $\boxed{aw}$ edit descriptors, where the first form automatically determines the necessary width to contain the string in the I/O list. The second form allows manual specification of the width but, unlike the similar mechanism for numbers, the value is not invalidated with asterisks if the string in the I/O list is larger than w. Instead, the non-fitting part of the string on the right-hand side is simply truncated. Alternatively, if w is larger than the length of the string in the I/O list, the string will be right-justified.

All data edit descriptors can be preceded by a positive integer, when more values for which the same format is appropriate appear in the I/O list. This is particularly useful when working with arrays, as we will illustrate in Sect. 2.6.5.

Control edit descriptors: these do not assist in data I/O, but allow instructing the I/O system to perform other operations related to the alignment of output. We only discuss how to insert spaces and start a new line here (see Metcalf et al. [10] for other details).

To insert spaces in output, use the $\boxed{nx}$ edit descriptor, where n represents the number of spaces to be inserted. Similarly, to start a new record (line) without issuing another output statement, use the $\boxed{n/}$ edit descriptor, where n represents the number

---

[34] However, unlike `integer`, the value of d remains important even in this case, since truncation is usually inevitable when converting floating-point binary numbers to the decimal base.

of records to be marked as complete.[35] The following program uses these ideas, to print three `character`-strings and a `logical` value, where the first two strings are separated by two spaces, and three empty lines separate the second from the third string:

```fortran
program mixing_edit_descriptors1
  implicit none
  logical :: convergenceStat = .true.

  print '(a, 2x, a, 4/, a, 11)', &
      "Simulation","finished.", &
      "Convergence status = ", &
      convergenceStat
end program mixing_edit_descriptors1
```

Listing 2.11 | `src/Chapter2/mixing_edit_descriptors1.f90`

NOTE
The idea of counts (also known as "repeat counts") in front of edit descriptors is actually more general, since these can also appear in front of data edit descriptors (e.g. '(10i0)'), or even in front of groups of edit descriptors, surrounded by parentheses (e.g. '(5(f8.2, x))'). These are useful mostly when working with arrays, therefore we discuss them in more detail in Sect. 2.6.5.

Character string edit descriptors: we already presented cases when character strings were already present in the format specification itself. These are permitted (but only for output), and can be combined with other types of descriptors, leading to output statements like in the next program:

```fortran
program mixing_edit_descriptors2
  implicit none
  integer :: myInt = 213
  real :: myReal = 3.14

  print '("An integer: ", i3, /,"A real: ", f0.2)', &
      myInt, myReal
end program mixing_edit_descriptors2
```

Listing 2.12 | `src/Chapter2/mixing_edit_descriptors2.f90`

which should look more natural to C programmers.[36]

Managing format specifications: In the examples presented so far, we have written the format specification next to the I/O statements, as a string constant. This can be inconvenient in several situations, for example:

- when the same format specification needs to be reused for many I/O statements (here, the approach we have illustrated so far would lead to code duplication—always a red flag)

---

[35] Note that, if the current record is not empty, the number of empty records inserted by such an edit descriptor is $n - 1$.

[36] In C, the equivalent statement would be: `printf("An integer: %3d\nA real: %0.2f\n", anInt, aFloat);`

- when some facts about the format are not known until actual program execution (here, the string constant would impose switching between various hard-coded formats)

Fortunately, format specifications can also be non-constant strings, constructed dynamically at runtime. This can be used to address both issues above.[37] The following program illustrates how such a specification can be used for multiple output statements:

```
program string_variable_as_format_spec
  implicit none
  integer :: a = 1, b = 2, c = 3
  real :: d = 3.1, e = 2.2, f = 1.3
  ! format-specifier to be reused (could also use deffered-length)
  character(len=*), parameter :: outputFormat ='(i0, 3x, f0.10)'

  print outputFormat, a, d
  print outputFormat, b, e
  print outputFormat, c, f
end program string_variable_as_format_spec
```

Listing 2.13 | src/Chapter2/string_variable_as_format_spec.f90

2.4.3 Information Pathways: Customizing I/O Channels

The I/O statements discussed in the previous sections used the standard I/O channels: we always assumed that input is directed from the keyboard, and output is appearing on the screen. However, Fortran also allows the use of other channels (files or even character-strings), as will be discussed in this section.

Any I/O-channel (e.g. keyboard, screen, or a file on disk) is mapped to a unit. To distinguish between the various channels, each unit is identified by an integer unit-number, which is either

- selected by the platform (usually "5" represents standard-input from keyboard, and "6" standard-output to screen), or
- specified by the programmer (examples of this shown later).

General I/O-statements: The simplified forms of the I/O statements discussed previously (print and read) do not support customization of I/O channels. To gain more control, the general I/O statements (write and read[38]) need to be used, which we introduce below:

```
! general form of input statement
read ([unit=]u [,fmt=fm1] [,iostat=statCode] [,err=lbl1] [,end=lbl2]) &
  [inputList]
! general form of output statement
write([unit=]u [,fmt=fm1] [,iostat=statCode] [,err=lbl1]) [outputList]
```

---

[37] There is also the option to use format-statements, as we mentioned previously. However, their usefulness is limited to the first issue, which is why we chose not to describe them—see Metcalf et al. [10] for details.

[38] The general input statement has the same name as the simplified form, but observe the other differences.

As usual, the square brackets denote optional items. The unit-number (u) and the format specification (*fml*) are the only mandatory items (optionally, they can be preceded by $\boxed{\text{unit=}}$ and $\boxed{\text{fmt=}}$ respectively, to improve readability). Both of these items can be set to $\boxed{*}$, to recover the particular forms of I/O we already presented:

```fortran
program general_can_recover_special_io
   implicit none
   integer :: anInteger

   ! special forms, default formatting...
   read  *, anInteger                        ! input
   print *,"You entered: ", anInteger ! output

   ! ...equivalent general forms, default formatting
   read (*, *) anInteger                         ! input
   write(*, *)"You entered: ", anInteger ! output

   ! special forms, custom formatting...
   read '(i20)', anInteger                      ! input
   print'("You entered: ", i0)', anInteger ! output

   ! ...equivalent general forms, custom formatting
   read  (*,'(i20)') anInteger                    ! input
   write (*,'("You entered: ", i0)') anInteger ! output
end program general_can_recover_special_io
```

Listing 2.14 $\boxed{\texttt{src/Chapter2/general\_can\_recover\_special\_io.f90}}$

Expecting the unexpected: exception handling The remaining (optional) parameters in the general I/O-statements (which we named in the examples above *statCode*, *lbl1* and *lbl2* for read / *statCode* and *lbl1* for write) help the program recover from various exceptional conditions. Since the success of these I/O statements depends on properties of data channels usually beyond the control of our programs, many things can go wrong, without being a program bug. For example, when trying to read from a file, the file may not exist, or our program may not have permission to read from it. Similarly, the program may try to write to a file for which it has no write-permission, or there may not be sufficient space on the external device to contain the output data.

If the error-handling parameters are omitted, any problems encountered during the I/O operations will cause the program to crash, which is acceptable for test programs. However, for "industrial-strength" programs that will be run by many users, it is a good idea to put these error-handling facilities to good use, for example to assist the users. The meaning of the optional parameters is summarized below:

- $\boxed{\texttt{iostat=}\textit{statCode}}$: here, *statCode* is an integer which will be set to a value representing the status of the I/O operation (following the Unix tradition, zero means "no error", while a non-zero value signals that an error occurred)
- $\boxed{\texttt{err=}\textit{lbl1}}$: *lbl1* is the label[39] of a statement (within the same (sub)program), to which the program will jump if an error occurred during the I/O statement

---

[39] In Fortran, every statement can be given a *label*, which is simply a positive integer (of at most 5 digits), written before the statement. These provide "bookmarks" within the code, allowing the program to "jump" to that statement when necessary—either transparently to the user (when the jump results from error handling), or explicitly (using the controversial go to statement). *Please note that explicit jumps with go to are strongly discouraged, as they can quickly make programs difficult to understand!*

- end=*lbl2* : *lbl2* is the label of a statement (within the same (sub)program), to which the program will jump if an "end-of-file" condition will be met (for the read-statement)

The following program illustrates how these extra arguments may be used:

```fortran
program read_with_error_recovery
  implicit none
  integer :: statCode=0, x

  ! The safeguarded READ-statement
  read(unit=*, fmt=*, iostat=statCode, err=123, end=124) x
  print'(a, 1x, i0)',"Received number", x
  ! Normal program termination-point, when no exceptions occur
  stop

123 write(*,'(a, 1x, i0)') &
      "READ encountered an ERROR! iostat =", statCode
  ! can insert here code to recover from error, if possible...
  stop
124 write(*,'(a, 1x, i0)') &
      "READ encountered an end-of-file! iostat =", statCode
  ! can insert here code to recover from error, if possible...
  stop
end program read_with_error_recovery
```

Listing 2.15 | src/Chapter2/read_with_error_recovery.f90

Exercise 3 (*Testing error recovery*) Compile the program listed above, and try providing different types of input data, to test how the error-handling mechanism works.

Hints: try providing (a) a valid integer-value, (b) a string and (c) an end-of-file character (on Unix: type CTRL+d).

The three phases of I/O: Working with external data channels in Fortran implies the following sequence of phases:

1. **establishing the link**: before the I/O system can use a unit, a link needs to be established and a unique unit-number assigned. For standard I/O (keyboard/screen), the channels are pre-connected by the Fortran runtime system, without any intervention from the programmer.

   However, for all other cases the link has to be established explicitly, with the open-statement. From the programmer's point of view, the most important effect of this statement is to associate a unit-number to the actual data channel. This number is necessary for the next steps (e.g. when the actual I/O takes place). Currently, there are two methods for performing this association:

   a. Until Fortran 2003, the programmer was responsible for explicitly selecting a positive integer-value for the unit-number. For working with ASCII files,[40] the open-statement would then commonly look like:

---

[40] Creating "binary" files is also possible, but we avoid discussing this, in favor of another format which is more appropriate in ESS, i.e., netCDF (see Sect. 5.2.2).

```
open([unit=]unitNum [, file=fileName] &
     [, status=statusString] [, action=actionString] &
     [, iostat=statCode] [, err=labelErrorHandling] &
     )
```

where:

- $\boxed{\texttt{unitNum}}$ is a positive `integer` variable/constant, assigned by the programmer. This will be used by the actual I/O statements.

- $\boxed{\texttt{fileName}}$ is a `character`-string, representing the actual name of the file in the file system.[41] This can be omitted only when $\boxed{\texttt{statusString}}$ $\boxed{\texttt{=="scratch"}}$ (which is useful for creating temporary files, managed by the system, and usually deleted when the program terminates).

- $\boxed{\texttt{statusString}}$ is one of the following strings: "old", "new", "replace", "scratch" or "unknown" (=default). This can be used to enforce some assumptions related to the status of the file prior to opening it.

- $\boxed{\texttt{actionString}}$ is one of the strings: "read", "write" or "readwrite". This is useful for limiting the set of I/O statements that can be used with the `unit`, which can help prevent bugs.

- $\boxed{\texttt{statCode}}$ and $\boxed{\texttt{labelErrorHandling}}$ have the same roles as `statCode` and `lbl2` in the preceding discussion on error-handling.

The following listing presents some examples:

```
10   integer :: statCode
11   real :: windUx=1.0, windUy=2.0, pressure=3.0
12
13   ! assuming file "wind.dat" exists, open it for reading, selecting
14   ! the value of 20 as unit-id; no error-handling
15   open(unit=20, file="wind.dat", status="old", action="read")
16
17   ! open file "pressure.dat" for writing (creating it if it does not
18   ! exist, or deleting and re-creating it if it exists), selecting
19   ! the value of 21 as unit-id; place in variable 'statCode' the
20   ! result of the open-operation
21   open(unit=21, file="pressure.dat", status="replace", &
22        action="write", iostat=statCode)
23
24   ! open a scratch-file, for storing some intermediate-result (which
25   ! we need to read later), that would be too large to keep in memory;
26   ! no error-handling
27   open(unit=22, status="scratch", action="readwrite")
```

Listing 2.16 $\boxed{\texttt{src/Chapter2/file\_io\_demo\_manual\_unit\_numbers.}}$ $\boxed{\texttt{f90}}$ (excerpt)

b. Requiring the programmer to manually manage the `unit`-numbers (the "magic" numbers 20, 21, and 22 in the listing above) is inconvenient, especially for large projects. Fortunately, since Fortran 2008, it is possible to ask the runtime system to automatically provide a suitable `unit`-number, so that clashes with any other open links are avoided. The syntax for the `open`-statement is similar to the one previously shown, except that we need to replace $\boxed{\texttt{[unit=]unitNum}}$ with $\boxed{\texttt{[newunit=]unitVariable}}$:

---

[41] Note that there might be some system-dependent restrictions on what constitutes a valid filename.

```
open([newunit=]unitVariable [, file=fileName] &
     [, status=statusString] [, action=actionString] &
     [, iostat=statCode] [, err=labelErrorHandling] &
     )
```

Note that, with this new method, it is not possible anymore to use constants for the `newunit`-value—only `integer` variables are accepted. This is because, when the `open`-statement is invoked, the runtime system will need to update `unitVariable`.[42]

With this new method, the examples presented above can be re-written as:

```
13   integer :: statCode, windFileID, pressureFileID, scratchFileID
14   real :: windUx=1.0, windUy=2.0, pressure=3.0
15   ! assuming file"wind.dat"exists, open it for reading, and store an
16   ! (automatically-acquired) unit-number in variable'windFileID'; no
17   ! error-handling
18   open(newunit=windFileID, file="wind.dat", status="old", &
19       action="read")
20
21   ! open file"pressure.dat"for writing (creating it if it does not
22   ! exist, or deleting and re-creating it if it exists), while storing
23   ! the (automatically-acquired) unit-number in variable'pressureFileID';
24   ! place in variable'statCode' the result of the open-operation
25   open(newunit=pressureFileID, file="pressure.dat", status="replace", &
26       action="write", iostat=statCode)
27
28   ! open a scratch-file, storing the (automatically-acquired) unit-number
29   ! in variable'scratchFileID'; no error-handling
30   open(newunit=scratchFileID, status="scratch", action="readwrite")
```

Listing 2.17 `src/Chapter2/file_io_auto_manual_unit_numbers.`
`f90` (excerpt)

Good practice
Due to its convenience, we recommend to use this second method (using `newunit`) when opening files. We also rely on this technique in the later examples for this book (especially in Chap. 4).

2. **actual I/O calls**: the second phase corresponds to issuing the actual I/O-statements, for the data we want to read or write. We discussed this in the previous sections; the only change necessary for file I/O is that the $\boxed{*}$ used until now for the `unit`-id needs to be replaced by the appropriate variable, as associated in advance within the `open`-statement. For example (continuing the example from the previous listing):

```
32   ! ... some code to compute pressure ...
33   read(windFileID, *) windUx, windUy
34
35   ! display on-screen the values read from the"wind.dat"-file
36   write(*,'("windUx =", 1x, f0.8, 2x,"windUy =", 1x, f0.8)') &
37       windUx, windUy
```

---

[42] The standard specifies that a negative value (but different from $\boxed{-1}$, which signals an error) will be chosen for `unitVariable`, to avoid clashes with any existing code that uses the previous method of assigning `unit`-numbers, where positive numbers had to be used.

```
38
39      ! write to scratch-file (here, only for illustration-purpose; this makes
40      ! more sense if 'pressure' is a large array, which we would want to modify,
41      ! or deallocate afterwards, to save memory)
42      write(scratchFileID,'(f10.6)') pressure ! write to scratch
43      ! re-position file cursor at beginning of the scratch-file
44      rewind scratchFileID
45      ! ... after some time, re-load the 'pressure'-data from the scratch-file
46      read(scratchFileID,'(f10.6)') pressure
47
48      ! write final data to "pressure.dat"-file
49      write(pressureFileID,'(f10.6)') pressure*2
```

Listing 2.18 `src/Chapter2/file_io_auto_manual_unit_numbers.`
`f90` (excerpt)

3. **closing the link**: unlike the first phase (establishing the link), the system will automatically close the link to any active `unit`, if the program completes normally. It is, however, still recommended for the programmer to perform this step manually, to avoid losing data in case an exception occurs.[43] To terminate the link to a unit, the `close`-statement can be used:

```
close([unit=]unitNum [, status=statusString]
      [, iostat=statCode] [, err=labelErrorHandling]
      )
```

Like for the `open`-statement, `unitNum` is mandatory, but some additional (optional) parameters are also supported:

- `statusString` can be either "keep" (=default, if the unit does not correspond to a scratch file) or "delete" (=required value for scratch files)
- `statCode` and `labelErrorHandling` can be used for error-handling, like for the `open`-statement

For example, the files opened in the previous listings can be closed with:

```
52      close(windFileID); close(pressureFileID); close(scratchFileID)
```

Listing 2.19 `src/Chapter2/file_io_auto_manual_unit_numbers.`
`f90` (excerpt)

Internal files: In addition to units, the general I/O statements in Fortran can also operate on internal files (which are simply buffers, stored as strings or arrays of strings).[44]

Internal files are similar, in a sense, to the scratch files that we described earlier, since they are normally used for temporarily holding data which need to be manipulated at a later stage of the program's execution. However, because they are resident in

---

[43] Such data loss can occur when writing to files, since most platforms use buffering mechanisms for temporarily storing output data, to compensate for the slow speed of the permanent storage devices (e.g. disks).

[44] Strictly speaking, these do not form true I/O operations (the buffers are still memory areas associated with the program, so no external system is involved), but it is convenient to treat them as such (as done for the equivalent `stringstream` class in C++).

memory, they are usable only for smaller amounts of data. One application of internal files is type conversion between numbers and strings—for example, to dynamically construct names for the output files of an iterative model, at each time step.[45] One approach to achieve this is shown in the listing below:

```fortran
program timestep_filename_construction
  implicit none
  character(40) :: auxString ! internal file (=string)
  integer :: i, numTimesteps = 10, speedFileID

  ! do is for looping over an integer interval (discussed soon)
  do i=1, numTimesteps
    ! write timestep into auxString
    write(auxString,'(i0)') i
    ! open file for writing, with custom filename
    open(newunit=speedFileID, &
         file="speed_"// trim(adjustl(auxString)) //".dat", &
         action="write")

    ! here, we would have model-code, for computing the speed and writing
    ! it to file...

    close(speedFileID)
  end do
end program timestep_filename_construction
```

Listing 2.20 | src/Chapter2/timestep_filename_construction.f90

Non-advancing I/O: We illustrated towards the end of Sect. 2.4.1 how, unlike other languages, Fortran automatically advances the file-position with each I/O statement, to the beginning of the next record. However, this can be turned off for a particular I/O-statement, by setting the optional control specification advance to "no" (default value is "yes"). This is often used when data is requested from the user, in which case it is desirable to have the prompt and the user input on the same line. We already used this technique, in Listings 2.9 and 2.10.

2.4.4 The Need for More Advanced I/O Facilities

So far, we discussed some basic forms of I/O, which are useful in common practice. However, these approaches do not scale well to the data throughput of state of the art ESS models (currently, in the terrabyte range for high-resolution models with global coverage). Text ("formatted") files are ineffective for handling such amounts of data, since each character in the file still occupies a full byte. If we imagine a very simple file which only contains the number 13, the ASCII-representation will occupy $2\,bytes = 16\,bits$. In addition, to mark the end of each record, a newline character (Unix) or carriage-return + newline (Windows) needs to be added for every row in the file. Thus, the total space requirement for storing our number in a file will be of $3\,bytes$ on Unix, and $4\,bytes$ on Windows systems, respectively.

---

[45] Here, we imply there is one output file for each time step, to illustrate the idea. Note, however, that this may not always be a good approach. In particular, when the number of time steps is large, it is more convenient to write several time steps in each file (this is supported by the netCDF-format, which we will describe in Sect. 5.2.2).

Alternatively, if we choose to store the data directly in binary form, 4 *bits* would already be sufficient in theory to represent the number 13 (however, this is half of the smallest unit of storage—on most systems, the file would finally occupy 1 *byte*). These calculations illustrate that there is a large potential for reducing the final size of the files, even without advanced compression algorithms, just by storing data in the binary format instead of the ASCII representation. Other advantages include:

- *less CPU-time spent for I/O operations*: the conversion to/from ASCII also increases the execution time of the program, by an amount that can become comparable to the time spent for actual computations
- *approximation errors*: especially when working with floating-point data, approximation errors can be introduced each time a conversion between binary and ASCII representations takes place

While the benefits of binary storage are significant, it does have the problem that interpretation of data is made more difficult.[46] The importance of this cannot be overstated, which is why *it is not recommended* to use the binary format directly in most cases: a much more convenient solution in ESS is to use the `netCDF` data format, which allows efficient storage in a platform-independent way. We cover this topic later, in Sect. 5.2.2, after introducing some more language features.

2.5 Program Flow-Control Elements (`if`, `case`, Loops, etc.)

Most programs shown so far consisted of instructions that were executed in sequence. However, in real applications it is often necessary to break this ordering, as some blocks of instructions may need to be executed (once or several times) only when certain conditions are met. The generic name for such constructs is (*program*) *flow-control*, and Fortran has several of them, as we discuss in this section.

Style recommendation: In the examples below, we indent each block of program instructions, to clearly reflect situations when their execution is conditioned by a certain flow-control construct. Indentation is not required by the language (the compiler eventually removes whitespace anyway), but it *greatly* improves the clarity of the code, especially when multiple flow-control constructs are nested. *We highly recommend this practice.*

2.5.1 `if` Construct

The simplest form of flow-control can be achieved with the `if`-statement which, in its most basic form, executes a block of code only when a certain scalar logical condition is satisfied. This is illustrated by the following program, which asks for a number, and informs the user in case it is odd:

---

[46] Various technicalities (such as platform dependence of the internal, bit-level representation of the same data) can make the data transfer nontrivial for binary data.

```
program number_is_odd
   implicit none
   integer :: inputNum

   write(*,'(a)', advance="no")"Enter an integer number:  "
   read(*, *) inputNum
   ! NOTE: mod is an intrinsic function, returning the remainder
   ! of dividing first argument by the second one (both integers)
   if( mod(inputNum, 2) /= 0 ) then
      write(*,'(i0, a)') inputNum," is odd"
   end if
end program number_is_odd
```

Listing 2.21 | src/Chapter2/number_is_odd.f90

In this case (when there is only one branch in the if), the corresponding code can be made even more compact, on a single line[47]:

```
if( mod(num, 2) /= 0 ) write(*,'(i0, a)') num," is odd"
```

We may wish to extend the previous example, such that a message is printed also when the number is even. This can also be achieved with if, which supports an (optional) else-branch:

```
if( mod(num, 2) /= 0 ) then
   write(*,'(i0, a)') num," is odd"
else
   write(*,'(i0, a)') num," is even"
end if
```

Sometimes, if the primary logical condition of the if-construct is .false. , we may need to perform additional tests. This is still possible using if only, in the most general form of the construct, which introduces else if branches:

```
if( <logical_condition1> ) then
   ! block of statements for"then"
else if( <logical_condition2> ) then
   ! block of statements for first"else if"branch
else if( <logical_condition3> ) then
   ! block of statements for second"else if"branch
else
   ! block of statements if all logical conditions
   ! evaluate to .false.
end if
```

To illustrate, assume that we need to extend our previous example such that, when the number is even, we inform the user if it is zero. This can be implemented as in:

```
if( mod(num, 2) /= 0 ) then
   write(*,'(i0, a)') num," is odd"
! num is odd, now check if it is zero
else if( num == 0 ) then
   write(*,'(i0, a)') num," is zero"
! default,"catch-all"branch, if all tests fail
else
   write(*,'(i0, a)') num," is non-zero and even"
end if
```

---

[47] Note that the keywords then and end if do not appear in the compact form.

Other constructs (including other `if`-statements) can appear within each of the branches of the conditional.[48] It is recommended to moderate this practice (since it can easily lead to code that is hard to follow), but sometimes it cannot be avoided. In such cases, proper indentation becomes crucial. Also helpful is the fact that Fortran allows `if`s (as well as the rest of the flow-control constructs) to be named, to make it clear to which construct a certain branch belongs; when names are used, the branches need to bear the same name as the parent construct. This is illustrated in the following (artificial and a little extreme) example, which asks the user for a 3-letter string, and then reports the corresponding northern hemisphere season[49]:

```fortran
program season_many_nested_ifs
  implicit none
  character(len=30) :: line
  write(*,'(a)', advance="no")"Enter 3-letter season acronym: "
  read(*,'(a)') line

  if( len_trim(line) == 3 ) then
    winter: if( trim(line) =="djf") then
      write(*,'(a)')"Season is: winter"
    else if( trim(line) =="DJF") then winter
      write(*,'(a)')"Season is: winter"
    else winter
      spring: if( trim(line) =="mam") then
        write(*,'(a)')"Season is: spring"
      else if( trim(line) =="MAM") then spring
        write(*,'(a)')"Season is: spring"
      else spring
        summer: if( trim(line) =="jja") then
          write(*,'(a)')"Season is: summer"
        else if( trim(line) =="JJA") then summer
          write(*,'(a)')"Season is: summer"
        else summer
          autumn: if( trim(line) =="son") then
            write(*,'(a)')"Season is: autumn"
          else if( trim(line) =="SON") then autumn
            write(*,'(a)')"Season is: autumn"
          else autumn
            write(*,'(5a)') &
              '"', trim(line), '"'," is not a valid acronym", &
              " for a season!"
          end if autumn
        end if summer
      end if spring
    end if winter
  else
    write(*,'(5a)') &
      '"', trim(line), '"'," is cannot be a valid acronym", &
      " for a season, because it does not have 3 characters!"
  end if
end program season_many_nested_ifs
```

Listing 2.22 `src/Chapter2/season_many_nested_ifs.f90`

Note that, while indentation and naming of constructs are helpful, the resulting code still looks complex, which is why we do not recommend including such extreme forms of nesting in real applications. For the current example, there is a way to simplify the logic using the `case`-construct, discussed next.

Note on spacing: In Fortran, several keywords (especially for marking the termination of a flow-control construct) can be written *with* or, equivalently, *without* spaces

---

[48] The process is called *nesting*. When used, nesting has to be complete, in the sense that the "parent"-construct must include the "child"-construct entirely (it is not allowed to have only partial overlap between the two).

[49] This is a common convention in ESS, where DJF = winter, MAM = spring, JJA = summer, and SON = autumn (for the northern hemisphere). The acronyms are obtained by joining the first letters of the months in each season.

in between. For example, `endif` is equivalent to `end if`, and `enddo` (discussed later)—to `end do`. This is more a matter of developer preferences.

2.5.2 *case Construct*

Another flow-control construct is `case`, which allows comparing an expression (of `logical`, `integer`, or `character` type) against different values and ranges of values. The general syntax for it is:

```
select case( <expression> )
case ( <match_list1> )
   ! block of statements when expression evaluates to
   ! a value in match_list1

case ( <match_list2> )
   ! block of statements when expression evaluates to
   ! a value in match_list2

! ... (other cases)

case default
   ! block of statements when no other match was found
   ! ("catch-all"case)
end select
```

Unlike the `if`-construct, where multiple expressions could be evaluated by adding `else if`-branches, `case` only evaluates one expression, and afterwards tries to match this against each of the cases. To avoid ambiguities, the patterns in the different match-lists are *not* allowed to overlap.

Note that only (literal) constants are allowed in each match-list. An interesting feature related to the match-list is that ranges of values are allowed (for types `integer` and `character`). Furthermore, values and ranges can be combined freely. This is shown in the following example, which reads a character, and tests if it is a vowel (assuming the English alphabet):

```
program vowel_or_consonant_select_case
   implicit none
   character :: letter

   write(*,'(a)', advance="no") &
      "Type a letter of the English alphabet: "
   read(*,'(a1)') letter

   select case( letter )
   case ('a','e','i','o','u', &
      'A','E','I','O','U')
      write(*,'(4a)')'"', letter, '"'," is a vowel"
      ! note below: match-list consists of values,
      ! as well as value-ranges
   case ('b':'d','f','g','h','j':'n','p':'t','v':'z', &
      'B':'D','F','G','H','J':'N','P':'T','V':'Z')
      write(*,'(4a)')'"', letter, '"'," is a consonant"
   case default
      write(*,'(4a)')'"', letter, '"'," is not a letter!"
   end select
end program vowel_or_consonant_select_case
```

Listing 2.23 | src/Chapter2/vowel_or_consonant_select_case.f90

For specifying ranges of values, it is even allowed to omit the lower or the higher bound (but not both), which allows ranges to extend to the smallest (negative) and largest (positive) representable `integer-value`.[50] This is used in the next code listing, which asks the user to enter an integer value, and checks if the number is a valid index for a calendar month:

```fortran
program check_month_index_select_case_partial_ranges
  implicit none
  integer :: month
  write(*,'(a)', advance="no")"Enter an integer-value: "
  read(*, *) month

  ! check if month is valid month-index, with partial
  ! (semi-open) ranges in a select-case construct
  select case( month )
  case ( :0, 13: )
     write(*,'(a, i0, a)')"error: ", &
         month," is not a valid month-index"
  case default
     write(*,'(a, i0, a)')"ok: ", month, &
         " is a valid month-index"
  end select
end program check_month_index_select_case_partial_ranges
```

Listing 2.24 `src/Chapter2/check_month_index_select_case_par-tial_ranges.f90`

Using the `case`-construct can lead to great simplifications of what would otherwise be complex, nested `if`-contraptions. For example, the season-acronym matching program, could be re-written as:

```fortran
program season_select_case
  implicit none
  character(len=30) :: line

  write(*,'(a)', advance="no")"Enter 3-letter season acronym: "
  read(*,'(a)') line

  if( len_trim(line) == 3 ) then
    season_match: select case( trim(line) )
    case ("djf","DJF") season_match
       write(*,'(a)')"Season is: winter"
    case ("mam","MAM") season_match
       write(*,'(a)')"Season is: spring"
    case ("jja","JJA") season_match
       write(*,'(a)')"Season is: summer"
    case ("son","SON") season_match
       write(*,'(a)')"Season is: autumn"
    case default season_match
       write(*,'(5a)') &
           '"', trim(line), '"'," is not a valid acronym", &
           " for a season!"
    end select season_match
  else
    write(*,'(5a)') &
        '"', trim(line), '"'," is cannot be a valid acronym", &
        " for a season, because it does not have 3 characters!"
  end if

end program season_select_case
```

Listing 2.25 `src/Chapter2/season_select_case.f90`

where we also demonstrated how to assign a name (in this example: `season_match`) to the `case`-construct.

---

[50] These are, in a sense, the discrete equivalents of $\pm\infty$.

2.5.3 *do Construct*

The flow-control constructs discussed so far (if and case) allow us to determine whether blocks of code need to be executed or not. Another pattern, which
is extremely important in modeling, is to execute certain blocks of code *repeatedly*,
until some termination criterion is satisfied. This pattern (also known as *iteration*) is
supported in Fortran through the do-construct, which we describe in this section.

The simplest form of iteration uses an integer-counter, as in the following
example:

```
integer :: i
do i=-15, 10
  ! block of statements, to be executed for each iteration
  write(*,'(i0)') i
end do
```

Here, the variable i is also known as the *loop counter*, and needs to be of integer
type. The numbers on *line 2* represent the lower (-15) and upper bound (10). For
each value in this range, the block of statements within the do-loop will be executed.
Within this block, the value of i can be *read* (e.g. it can appear in expressions), but
it *cannot* be *modified*.

2.5.3.1 Loop Counter Increment

By default, the loop counter is incremented by one at the end of each iteration. Fortran
also allows to specify a different increment, as a third number at the beginning of
the do-construct. This allows, for example, incrementing the loop counter in larger
steps, or even decrementing it, to scan the range of numbers backwards. For example:

```
! iterate from 0 to 100, in steps of 25
do i=0, 100, 25
  ! block of statements
end do
! iterate backward, from 8 to -8, in steps of 2
do i=8, -8, -2
  ! block of statements
end do
```

In our examples so far, we always used integral literals for the *start-*, *end-*, and
*increment-*values of the loop counter. However, the language also allows these to
be integer-variables, or even more complex expressions involving variables. In
such cases, the variables can be altered within the loop, but this has no influence
whatsoever on the progress of the loop, since only the initial values are used for
"planning" the loop. For example, in the following listing, the assignments on *lines
6* and *7* have no impact on the loop:

```fortran
program do_specified_with_expressions
  implicit none
  integer :: timeMax = 10, step = 1, i, numLoopTrips = 0

  do i=1, timeMax, step
    timeMax = timeMax / 2
    step = step * 2
    numLoopTrips = numLoopTrips + 1
    write(*,'(a, i0, a, /, 3(a, i0, /))') &
      "Loop body executed ", numLoopTrips," times", &
      "i = ", i, &
      "timeMax = ", timeMax, &
      "step = ", step
  end do

end program do_specified_with_expressions
```

Listing 2.26   `src/Chapter2/do_specified_with_expressions.f90`

Exercise 4 (*Practice with do-loops*) The equidistant cylindrical projection is one of the simplest methods to visualize the Earth surface in a plane. This projection maps meridians and parallels onto vertical and horizontal lines, respectively. However, this projection is not "equal area"—for example, axis-aligned rectangles (say, $10°$ latitude $\times 10°$ longitude) which have the same area on the map do not have equal areas in reality.

To quantify this effect, use a do-loop to evaluate areas of 9 such cells (with latitude bounds $[0°N, 10°N]$, $[10°N, 20°N]$, ...$[80°N, 90°N]$. How large is the area of the near-pole cell, relative to that of the near-equator cell (in percents)?

Hint: Assuming our vertical displacement is much smaller than the average Earth radius, a "cell" whose normal coincides with the local direction of gravity has an area given approximately by:

$$S_i = R_E^2 \left(\lambda^E - \lambda^W \right) \left(\sin \phi^N - \sin \phi^S \right),$$

where both latitudes ($\lambda^{\{E,W\}}$) and longitudes ($\phi^{\{S,N\}}$) are given in radians.

Exercise 5 (*Hypothetical potential density profile*) Assume the potential density profile for a rectangular box within the ocean is given by:

$$\sigma_\theta(y, z) = \left[0.9184 \left(-\sqrt{G(y, z)} + 1 \right)^2 + 0.9184 \arccos^2 \right.$$

$$\left. \left(\frac{1}{\sqrt{G(y, z)}} \left(2 - \frac{y}{H} \right) \right) + 26.57 \right] \text{kg/m}^3 \qquad (2.1)$$

$$G(y, z) = \left(2 - \frac{y}{H} \right)^2 + \left(0.1 + \frac{z}{H} \right)^2 \qquad (2.2)$$

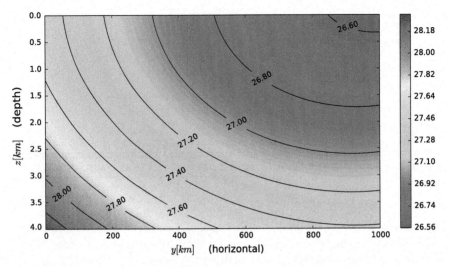

Fig. 2.1 Idealized profile of potential density (σ_θ), based on Eqs. (2.1)–(2.2)

where:
- $y \in [0, L]$, with: $L = 1000\,\text{km}$
- $z \in [0, H]$, with: $H = 4\,\text{km}$

This can be viewed as an idealized profile of the density structure in some part of the ocean (Fig. 2.1).

Assuming the extent along the x-axis (perpendicular to the figure) is of 100 km, compute the fraction of total volume occupied by water whose potential density matches the range typical for upper Labrador Sea Water (uLSW), which is:

$$\sigma_\theta^{\text{uLSW}} \in [27.68, 27.74]\,\text{kg m}^{-3}$$

2.5.3.2 Non-deterministic Loops

In practical applications, loops are not always deterministic.[51] Suppose we need to read successive data elements (e.g. a time series) from a file, for estimating the mean and the variance of the values. The steps of the algorithm are the same for each considered value, so it is natural to surround them by a loop construct. However, since the data resides in the external file, we may not know in advance how many values there are. Fortran accommodates such cases with the "endless" do construct, which looks like:

---

[51] "Non-deterministic" means, in this context, *not* (easily) *determined at compilation time*.

```
do
    ! block of statements
end do
```

This form truly has the tendency to run endlessly,[52] and it is the responsibility of the programmer to devise a suitable termination criterion, and to end the execution of the loop with the `exit`-statement. This is illustrated in the following listing, which demonstrates a way to solve the file-reading problem described above, where a suitable loop-termination criterion is that the end-of-file was reached while trying to read-in data:

```
1  program mean_and_standard_deviation_from_file
2    implicit none
3    integer :: statCode, numVals=0, inFileID
4    real :: mean=0.0, variance=0.0, sd=0.0, newValue, &
5           sumVals=0.0, sumValsSqr=0.0
6
7    ! open file for reading
8    open(newunit=inFileID, file="time_series.dat", action="read")
9
10   !"infinite"DO-loop, to read an unknown amount of data-values
11   data_reading_loop: do
12       read(inFileID, *, iostat=statCode) newValue
13       ! check if exception was raised during read-operation
14       if( statCode /= 0 ) then ! **TERMINATION-CRITERION for DO-loop**
15          exit data_reading_loop
16       else ! datum read successful
17          numVals = numVals + 1
18          sumVals = sumVals + newValue
19          sumValsSqr = sumValsSqr + newValue**2
20       end if
21   end do data_reading_loop
22
23   ! close file
24   close(inFileID)
25
26   ! evaluate mean (avoiding division by zero, when file is empty)
27   if( numVals > 0 ) mean = sumVals / numVals
28   ! evaluate 2nd central-moment (variance)
29   variance = (sumValsSqr - numVals*mean**2) / (numVals - 1)
30   ! evaluate standard-deviation from variance
31   sd = sqrt( variance )
32
33   write(*,'(2(a, f10.6))')"mean = ", mean, &
34        ", sd = ", sd
35 end program mean_and_standard_deviation_from_file
```

Listing 2.27 | `src/Chapter2/mean_and_standard_deviation_from_`
`file.f90`

where we used the fact that:

$$s\{X\} \equiv \sqrt{var\{X\}} = \sqrt{\frac{\sum_{i=1}^{N}(x_i - \bar{x})^2}{N-1}} = \cdots = \sqrt{\frac{1}{N-1}\left(\sum_{i=1}^{N}x_i^2 - N\bar{x}^2\right)}$$

where s is the unbiased estimator of the standard deviation, N is the number of samples, $\bar{x}$ is the estimated mean, and $x_{i\in[1...N]}$ corresponds to the individual samples.

If the loop that we wish to terminate is named, it is possible to provide this name to the `exit`-statement, to improve the clarity of the code. We illustrated this in the example above, although the value of this feature is more obvious when several loops are nested.

---

[52] At least, until the program is terminated forcibly.

2.5.3.3 Shortcutting Loops

Another pattern that occurs sometimes while working with loops is skipping over parts of the code within the loop's body, when certain conditions are met, *without leaving the loop.* For example, assume we are writing a program which converts a given number of seconds into a hierarchical representation (weeks, days, hours, minutes, and seconds). Clearly, the number of seconds provided by the user should be positive for the algorithm to work. If the user provides a negative integer, it does not make sense to try to find a hierarchical representation of the period; instead, it would be more useful to skip the rest of the code within the loop, and proceed to the next loop iteration directly, where the user has the opportunity to provide another input value. This type of behavior is supported in Fortran, using the `cycle [loop_name]` [53] command, as illustrated in the following example:

```fortran
program do_loop_using_cycle
   implicit none
   integer, parameter :: SEC_IN_MIN = 60, &
         SEC_IN_HOUR = 60*SEC_IN_MIN, & ! 60 minutes in hour
         SEC_IN_DAY = 24*SEC_IN_HOUR, & ! 24 hours in a day
         SEC_IN_WEEK = 7*SEC_IN_DAY ! 7 days in a week
   integer :: secIn, weeks, days, hours, minutes, sec

   do
      write(*,'(/, a)', advance="no") & !'/' adds newline, for separation
            "Enter number of seconds (or 0 to exit the program): "
      read(*, *) secIn

      if( secIn == 0 ) then ! loop-termination criterion
            exit
      else if( secIn < 0 ) then ! skipping criterion
            write(*,'(a)')"Error: number of seconds should be"// &
               " positive. Try again!"
            cycle ! ** calculation skipped with CYCLE **
      end if

      ! calculation using the value
      sec = secIn ! backup value
      weeks = sec / SEC_IN_WEEK;  sec = mod(sec, SEC_IN_WEEK)
      days = sec / SEC_IN_DAY;    sec = mod(sec, SEC_IN_DAY)
      hours = sec / SEC_IN_HOUR;  sec = mod(sec, SEC_IN_HOUR)
      minutes = sec / SEC_IN_MIN; sec = mod(sec, SEC_IN_MIN)
      ! display final hierarchy
      write(*,'(6(i0, a))') secIn,"s = { ", &
            weeks," weeks, ", days," days, ", &
            hours," hours, ", minutes," minutes, ", &
            sec," seconds }"
   end do
end program do_loop_using_cycle
```

Listing 2.28 `src/Chapter2/do_loop_using_cycle.f90`

Nesting of loops is another very common practice in ESS modeling, naturally occurring from the discretization of space and time. Another example of loop nesting occurs in linear algebra, for example matrix multiplication or transposition.

---

[53] `loop_name` is an optional name, which allows to clarify to which loop the `cycle`-command should be applied, in case of multiple nested do-loops.

Exercise 6 (*Zero-padded numbers in filenames*) The program in Listing 2.20 produced filenames in which the numeric portion had a variable width. This may prevent some post-processing tools from correctly identifying the order of the files.

Extend the program, so that the numeric portion in filenames has a constant width (with zero-padding), which is calculated based on the value of num_timesteps.

Hints: if num_timesteps is zero, the required number of digits is obviously one; for the other cases, you can use the expression aint(log10(real(num_timesteps))) + 1 (we assume num_timesteps >= 0). Also, you can use a second internal file, to construct the format for the statement where the i is written to aux_strng (since a dynamic minimum width of the integer field needs to be specified).

Exercise 7 (*Detecting kinds of numeric types on your platform*) We now have the tools to complete the discussion on kind-values (Sect. 2.3.4). Write a program that uses the intrinsic functions selected_int_kind and selected_real_kind to determine the variants of these two numeric types available on your platform.

Hints: For each type, search the parameter space with do-loops. For integer, iterate through values for requestedExponentRange in the interval [0, 45], and write to a file the (requestedExponentRange, obtained_kind)-pairs, as determined by your program. For real, use two nested do-loops, to iterate through values of requestedExponentRange in the interval [0, 5500], and values of requestedPrecision in the interval [0, 60], and write to another file the (requestedExponentRange, requestedPrecision, obtained_kind)-triplets. Visualize your results as a scatter plot for integer and a filled contour map for real (the results for our platform are shown in Figs. 2.2 and 2.3).

Exercise 8 (*Working with another platform*) Use the program developed for the previous exercise to test the kind-values for a different platform (hardware and/or compiler). Compare the results with those obtained in Exercise 7.

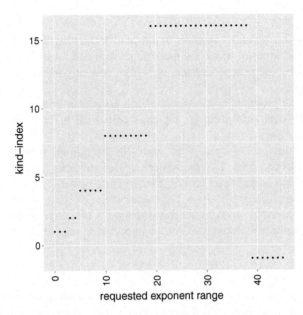

Fig. 2.2 `integer` kind indices as a function of requested exponent range (platform: `Linux`, 64 *bit*, `gfortran` compiler)

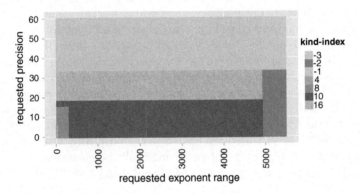

Fig. 2.3 `real` kind indices as a function of requested exponent range and requested precision (platform: `Linux`, 64 *bit*, `gfortran` compiler)

2.6 Arrays and Array Notation

So far, we used mostly scalar variables for representing entities in our example programs. This was sufficient, since the number of quantities was rather limited. However, in most applications (and ESS models in particular), the number of variables easily exceeds several millions, which is clearly not something that can be managed with scalars. There is, in fact, a distinct branch in computer science, dealing with *data structures*—methods of organizing data for various applications.[54] In this section, we focus on arrays, which are among the most basic, but also most popular data structures. In fact, arrays are so useful in scientific and engineering programs that a large part of the Fortran language is devoted to them.

An array is a compound object, which holds a set of elements. The elements can belong to any of the 5 intrinsic types already discussed, or even to derived types. An important constraint, however, is that all elements need to have the same type (and `kind`-parameter).

The second important aspect of arrays, besides the type of each element they store, is their shape. It is helpful to introduce some terms, which characterize this aspect for any given array:

- *rank* = number of dimensions of the array. "Dimensionality" in this context refers to the number of indices needed for uniquely specifying an element—similar to classification of tensors in mathematical physics.[55]
- *extent* = "width" along a particular dimension. Fortran arrays are *rectangular*, in the sense that the extent along each dimension is independent of the value of the indices along the other dimensions.[56] We will demonstrate later how the range for each index can be freely customized in Fortran, by specifying arbitrary (possibly negative) integers for the lower and upper bound. In this context, we have $\boxed{\texttt{extent == upper\_bound} - \texttt{lower\_bound} + 1}$.
- *shape* = 1*D*-array, each component of which represents an extent along a specific dimension.
- *size* = total number of scalar elements in the array (equals product of extents).

2.6.1 Declaring Arrays

Before working with arrays, we need to create them. This needs to be done explicitly in Fortran, and it implies *declaring* and *initializing* the arrays we want to use (second step is mandatory for constants, but highly recommended for modifiable arrays too).

In normal usage, there are two ways for declaring arrays, both of which require specification of the array *shape*. The first method uses the `dimension`-keyword, as in:

---

[54] Because the merits of a *data structure* can only be proven in the context of the *algorithms* applied on them, most references unify these two aspects (e.g. Mehlhorn and Sanders [9] or Cormen et al. [2]).

[55] At the risk of stating the obvious: this should not be confused with dimensionality of the physical space (if we store the components of a 3*D*-vector in an array, that array will have `rank==1`).

[56] So an entity with a more irregular shape, such as the set of non-zero elements of a lower-triangular matrix, needs to be stored indirectly when arrays are used.

```
! both X & Y are rank=1 arrays, with 16 elements each
real, dimension(16) :: X, Y
! A is a rank=3 array, with 520^3 elements
! up~to rank=15 is allowed in Fortran 2008 (was 7 in Fortran 90)
integer, dimension(520, 520, 520) :: A
```

The second declaration method is to specify the shape of the array after the variable name, as in:

```
! X is still a rank=1 array, but Y is a scalar real
real :: X(16), Y
! same effect as in previous declaration of A
integer :: A(520, 520, 520)
```

The numbers inside the shape specification actually represent the *upper bounds* for the indices along each dimension. An interesting feature in Fortran is that one can also specify *lower bounds*, to bring the code closer to the problem-domain:

```
real, dimension(-100:100) :: Z ! rank=1 array, with 201 elements
```

Notes

- *Unlike programming languages from the C-family, the value to which the lower bound defaults (when it is not specified) is 1 (**not 0**)!*
- *Although in the examples here we often specify the shape of the arrays using hard-coded integer values, it is highly recommended to use named integer constants[57] for this in real applications, which saves a lot of work when the size of the arrays needs to be changed (since only the value of the constant would need to be edited).*

2.6.2 Layout of Elements in Memory

We now turn our attention to a seemingly low-level detail which is, however, crucial for parts of our subsequent discussion: *given one of the array declarations above, how are the array elements actually arranged in the system's memory*[58]?

The memory can be viewed as a very large 1D sequence of bytes, where all the variables associated to our program are stored. For 1D-arrays, it is only natural to store the elements of the array contiguously in memory. Things are more complex for arrays of rank > 1, where an ordering convention (also known as "array element order") for the array elements needs to be adopted (effectively, defining a *mapping* from the tuple of coordinates in the array to a linear index in memory).

---

[57] This is achieved with the parameter-attribute.

[58] Here, we refer to the Virtual Memory subsystem, which includes mainly the *random-access memory* (RAM) and, less used nowadays, portions on secondary storage (e.g. hard-drives) which extend the apparent amount of memory available.

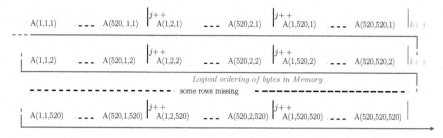

Fig. 2.4 Illustration of element ordering for a $3D$ array in Fortran. The *dashed horizontal black line* represents incrementing in the first dimension, the *black vertical lines*—incrementing in the second dimension, and the *vertical green lines*—incrementing in the third dimension. The *blue line* represents the logical ordering of bytes in memory. The figure was split into multiple rows, to fit in the page

---

NOTE
In Fortran, the array element order for elements of a multi-dimensional array is such that the earlier dimensions vary faster than the latter ones.[a]
This is exactly opposite to the corresponding convention in C *and* C++, *providing opportunities for bugs to appear while porting applications!*

---

[a] An alternative way to remember this is relative to how a matrix is stored: since the elements within a column are adjacent, Fortran (along with other languages like MATLAB and *GNU Octave* (octave)) is said to use *column-major order* (C and C++ use *row-major order*).

For example, the elements of the A-array declared earlier could be arranged in memory similarly to Fig. 2.4.

The array element order is important for understanding how several facilities of the language work with multi-dimensional arrays. It is also very relevant for application performance,[59] as illustrated in Exercise 9.

2.6.3 Selecting Array Elements

Since arrays group multiple elements, a crucial feature when working with them is the ability to select elements based on some pattern, which is usually dictated by a subtask of the algorithm to be implemented. Fortran supports many methods for

---

[59] This relates to the memory-hierarchy within modern systems. There are usually several layers of *cache*-memory (very fast, but with small capacity) between the CPU and RAM, to hide the relatively high latency for fetching data from RAM. Most caches implement a pre-fetching policy, and higher performance is achieved when the order in which array elements are processed is close to the array element order. Note that more details need to be considered, for performance-critical (sub)programs (for more information, see Hager and Wellein [5]).

outlining such selections. We illustrate these via examples below, assuming we want to overwrite some parts of an array. However, the same techniques apply for reading parts of an array, of course.

Given an array declaration like:

```
integer, parameter :: SZ_X=40, SZ_Y=80
! Note the use of named integer constants for specifying
! the shape of the array (recommended practice).
real, dimension(-SZ_X:SZ_X, -SZ_Y:SZ_Y) :: temperature
```

Fortran allows to select:

- **the entire array**: by simply specifying the array's name:

```
temperature = 0. ! scalar written to selection (=whole array)
```

- **a single element**: by specifying the array's name, followed, within parentheses, by a list of n **indices**[60] (where n is the rank of the array):

```
! scalar written to element (i=1, j=2)
temperature(1, 2) = 10.
```

- **a sub-array**: by specifying the array's name followed, within parentheses, by a list of n **ranges** (n = rank of the array, as before). A range, in this context, is an integer interval, with an optional step,[61] as in:

```
! scalar written to element (i=1, j=2)
temperature(-SZ_X:0, -SZ_Y:SZ_Y:2) = 20.
```

- **a list of elements**: by specifying the array's name followed, within parentheses, by one or more array(s) of $\boxed{\texttt{rank==1}}$ (we call these *selection arrays*). Each selection array represents a list of values for a corresponding dimension (so only one selection array is necessary when the source array is 1D, two when the source array is 2D, etc.). The elements of the source array which eventually become selected are those with the coordinate-tuples within the Cartesian product of the sets represented by the selection arrays. The next listing uses this procedure to select the corners of the $2D$-array `temperature`:

```
! only 4 elements are selected (Cartesian product):
! (-SZ_X, -SZ_Y), (-SZ_X, SZ_Y), (SZ_X, SZ_Y), (SZ_X, -SZ_Y)
temperature( [ -SZ_X, SZ_X ], [ -SZ_Y, SZ_Y ] ) = 30.
```

where we used the $\boxed{[}$ and $\boxed{]}$ tokens, to create arrays inline.[62] We will present more uses of this technique in the next section.

---

[60] The list of indices can also be provided as a $1D$-array of size n.

[61] Such ranges are very similar to what we illustrated previously for the do program-flow construct, except that in this case commas ($\boxed{,}$) need to be replaced by colons ($\boxed{:}$).

[62] This notation was introduced in Fortran 2003. Note that there is also an older (equivalent) notation, using the tokens $\boxed{(/}$ and $\boxed{/)}$.

> **NOTE**
> When an array selection is used for writing to an array, it is not recommended to have, in the selection arrays, elements which are repeated, since this can lead to attempts to write more than one value to the same array element.[a]
>
> ---
>
> [a] Some compilers may allow this without warnings, although the standard declares these as illegal. In any case, the behavior in such situations is likely platform-dependent, and the recommendation holds.

2.6.4 Writing Data into Arrays

As soon as an array is declared, a first concern, before using the values of the array elements in other statements, is to initialize those values. Unlike other languages, the Fortran standard does not make any guarantee regarding data initialization (such as setting them to zero), so explicit action is required from the programmer in this respect.

Values can be assigned to array elements using several mechanisms, to fit various scenarios. Just as for scalar variables, these assignments can be combined[63] with the declaration line, as a compact method of initialization (therefore, the techniques shown in this section apply to initialization, as well as to assignment).

An important notion when writing data to an array is *conformability*: two data entities are said to be conformable if they are arrays with the same shape, or if at least one of them is a scalar. When one entity is assigned to another one, they need to be conformable (this is also necessary when forming array expressions, as discussed later).

2.6.4.1 Writing a Constant Value

One of the simplest write operations is to assign a scalar value to an entire array (or an array section), in which case all elements (selected elements) will be set to that value:

```
! either: declaration, followed by assignment
! before the values are used
real, dimension(0:20) :: velocity
velocity = 0.
! or: initialization directly at declaration-time
real, dimension(0:20) :: velocity = 0.
```

---

[63] For array constants this is, naturally, required.

2.6.4.2 Writing Element-by-Element

Another form of writing into an array is the "lower-level" fashion, using element-based assignments, (optionally) combined with loops. This is the most flexible method and, perhaps, also the most intuitive. As a simple example, here is a more verbose (but logically equivalent) version of the assignment for the `velocity` array from the previous listing:

```
integer :: i
! element-based assignment (equivalent to: velocity = 0.)
do i=0,20
  velocity(i) = 0.
end do
```

Despite being conceptually straightforward, *we recommend avoiding this procedure* when possible, in favor of the ones discussed previously (writing a constant value), or next (writing another array(selection)). Still, this form is sometimes justified, for example when:

- the assignment does not follow an obvious pattern, or
- there is a definite performance advantage (proven by benchmarks) for using this method instead of the other ones.

2.6.4.3 Writing Another Array (Section)

An array (or array-section) can also be assigned to another array (or section), as long as the two entities are conformable. For example:

```
integer :: array1(-10:10), array2(0:20)
! ... some code to compute array2 ...
array1 = array2    ! whole-array assignment
```

Note that the arrays are conformable even if the lower and upper bounds of the array indices are different for the two arrays, as it was the case here (only the *shape* matters): after the assignment, `array1(-10)` == `array2(0)` == ... == `array1(10)` == `array2(20)`.

The use of array sections is illustrated in the following listing, which swaps the value of each odd element with that of the next even element[64]:

```
integer :: array3(1:20), tmpArray(1:10)
! ... some code to initialize array3 ...
tmpArray = array3(1:20:2)
array3(1:20:2) = array3(2:20:2)
array3(2:20:2) = tmpArray
```

---

[64] This assumes the lower bound for the index is odd, and that the upper one is even.

2.6.4.4 Array Constructors

We already mentioned that arrays can be initialized based on other arrays, but then one could ask how are the latter arrays to be initialized. Fortran has a special facility for this problem—the *array constructor*. This consists of a list of values, surrounded by square brackets.[65] A common use of this is to define a *constant* array (with the parameter-keyword), as in:

```
integer, dimension(3), parameter :: meshSize = [ 213, 170, 10 ]
real, dimension(0:8), parameter :: weights = [ 4./9., &
   1./9., 1./36., 1./9., 1./36., &
   1./9., 1./36., 1./9., 1./36. ]
```

The arrays defined with the constructor syntax can also be used directly in expressions (as long as they are conformable with the other components of the expression), as any other array, for example:

```
integer, dimension(10) :: xRange
xRange = [ 1, 2, 3, 4, 5, 6, 7, 8, 9, 10 ]
```

2.6.4.5 Patterns in Array Constructors: Implied-do Loops

A drawback of the weights and xRange examples above (using constructor syntax) is that they tend to be quite verbose. The *implied*-do *loops* were introduced in Fortran to solve this problem, when the values follow a well-defined pattern. They act as a convenient shorthand notation, with the general form:

```
! Note: the expression below needs to be embedded into an
! actual array constructor (see next examples).
( expr1, expr2, ..., indexVar = exprA, exprB [, exprC] )
```

where:

- indexVar is a named scalar variable of type integer (usually named i, j, etc.); note that the *scope* of this variable is restricted to the implied-do loop, so it will not affect the value of the variable if used in other parts of the program
- expr1, expr2, ...are expressions (not necessarily of integer type), which may or may not depend on indexVar
- exprA, exprB, and exprC are scalar expressions (of integer type), denoting the lower bound, upper bound and (optional) increment step for indexVar

To illustrate the implied-do loops, we use them to re-write the operations above (for weights and xRange) in a more compact (but otherwise equivalent) form:

---

[65] Or surrounded by the pre-Fortran 2003 tokens $(/$ and $/)$.

```
! index variable for implied-do still needs to be declared
integer :: i

xRange = [ (i, i=1,10) ] ! uses declaration above

real, dimension(0:8), parameter :: weights = &
   [ 4./9., (1./9., 1./36., i=1,4) ]
```

The implied-do loop is eventually expanded, such that the list { expr1, expr2,..., } is repeated for each value of the indexVar, using the appropriate value of the index variable for each repetition. For instance, in our second example above, the list {1./9., 1./36.} is repeated 4 times (and the value of the index variable is not used for computing any component).

2.6.4.6 Array Constructors for Multi-dimensional Arrays

So far, we only used array constructors for building 1D arrays. It is also possible, however, to construct multi-dimensional arrays, with a two-step procedure:

1. construct a 1D-array tmpArray
2. pass tmpArray to the intrinsic function reshape, to obtain a multi-dimensional array

In practice, the two steps are commonly combined into a single statement. The following example illustrates this, for constructing a 10×20 matrix, where each element $a_{i,j} = i * j$:

```
real, dimension(10, 20) :: a = reshape(    &
   source = [ ((i*j, i=1,10), j=1,20) ],   &
   shape  = [ 10, 20 ]                      &
)
```

where we also demonstrated the way in which implied-do loops can be *nested* (essentially, by replacing one or more of the expressions expr1, expr2, ..., discussed above by another implied-do loop).

In its basic form,[66] the reshape implicit function takes two arguments (denoted by optional keywords source and shape), both of them being 1D arrays, and where shape should be constant, and with non-negative elements.

The elements are read, in array element order, from the source-array, and written, also in array element order, to the result array.

2.6.5 I/O for Arrays

Just as we demonstrated in Sect. 2.4 for scalar variables, it is also essential to read/write (parts of) arrays from/to external devices. In principle, the same ideas could

---

[66] Additional arguments are supported, although not discussed here—see, e.g. Metcalf et al. [10].

be used, by simply treating individual array elements as scalar variables. However, there are several techniques related to array I/O which can simplify these operations. This section is devoted to these techniques.

2.6.5.1 Default Format ($\boxed{*}$)

Just as for scalar variables, it is possible to let the system choose a default format, as in:

```
1  integer :: i, j ! dummy indices
2  integer, dimension(2,3) :: inArray = 0
3
4  write(*,'(a)')"Enter array (2x3 values):"
5  read(*,*) inArray
6  write(*,'(a)')"You entered:"
7  write(*,*) inArray
```

The input (provided for *line 5* in the listing above) can be provided over multiple records—the system will keep reading new records, until the elements in the I/O-list (whole array in our case) are satisfied.

The appearance of the output (generated by *line 7*) is, as in the case of scalars, platform-dependent. This was merely an aesthetic issue for scalars, but in the case of arrays it actually poses a serious problem, since the topological information of the array is effectively *lost*[67] (the lines in the output will not correspond, in most cases, to recognizable features of the array, such as rows and columns for 2D arrays). In the particular case of the previous listing the 6 array elements would normally fit on a single line of output.

In the remainder of this section, we discuss several methods for producing higher-quality output. Related to this, we also illustrate several methods for specifying the format specification, ranging from verbose to compact.

2.6.5.2 Implied-do Loops in the I/O-List

A first problem with the write-statement at *line 7* in the previous listing is that, when an array appears in the I/O-list, the I/O-system will effectively expand it internally to a list of array elements, taken in the array element order. We know, based on the discussion at the beginning of this section, that for a 2D array this order is the transpose of what would be needed to output the elements (given that Fortran I/O is record-based). This can be solved by modifying the I/O-list, so that it contains an implied-do loop instead of the array, as follows:

```
write(*, *) ( ( inArray(i,j), j=1,3), i=1,2 )
```

---

[67] Strictly speaking, it is still possible to deduce the coordinates of a specific element in the output list, by counting its position, and then comparing this with the expected array element order; however, this can hardly be called productive use of the programmer's time.

2.6.5.3 List of Formats (Verbose)

The previous listing causes the two rows of the array to be written on the same line. To separate them, we need to control the appearance of the output, using a customized format specifier, as we illustrated before for scalars. A first option to achieve this is to specify a verbose list of edit descriptors, as in:

```
write(*,'(x, i0, x, i0, x, i0, /, x, i0, x, i0, x, i0)') &
   ( ( inArray(i,j), j=1,3), i=1,2 )
```

2.6.5.4 Repeat Counts

The previous statement causes the two rows of the matrix to appear on separate lines, as intended. However, the format specifier is quite verbose, and it would be impractical to write in this form if the matrix were to be larger. We mentioned below that Fortran allows repeat counts to be placed in front of edit descriptors, or groups of edit descriptors within parentheses. In the current case, this can be used to make the format descriptor more compact, by factoring the $\boxed{\text{x, i0}}$-pattern:

```
write(*,'(3(x, i0), /, 3(x, i0))') &
   ( ( inArray(i,j), j=1,3), i=1,2)
```

2.6.5.5 Recycling of Edit Descriptors

Finally, we notice that Fortran has a mechanism for "recycling" edit descriptors, so that there can be more elements in the I/O-list than edit descriptors in the output format. When the I/O-subsystem "runs out" of edit descriptors, a new line of output is started, and the format specifier is re-used for the next elements in the I/O-list. This is perfect for our current purposes, as the output format can be further simplified using this feature:

```
write(*,'(3(x, i0))') &
   ( ( inArray(i,j), j=1,3), i=1,2)
```

2.6.6 Array Expressions

We emphasized above the usefulness of working with whole arrays and array sections, instead of manually iterating through the array elements with loops. Fortran allows a similar high level of abstraction for representing computations, with array expressions. Specifically, most unary intrinsic functions and operators can take a whole array (or an array selection) as an argument, producing another array, with the same shape, through element-wise application of the operation. The same idea applies to binary operators, as long as the arguments are *conformable*. The following program uses these techniques to evaluate the functions $\sin(x)$ and $\sin(x)+\cos(x)/2$ on a regular grid, spanning the interval $[-\pi, \pi]$:

```
1  program array_expressions1
2     implicit none
3     integer, parameter :: N=100
4     real, parameter :: PI=3.1415
5     integer :: i
6     real, dimension(-N:N) :: &
7        xAxis = [ (i*(pi/N), i=-N,N) ], &
8        a = 0, b = 0
9
10    ! Compact array-expressions, using elemental functions.
11    ! a(i) == sin( xAxis(i) )
12    a = sin(xAxis)
13    ! b(i) == sin( xAxis(i) ) + cos( xAxis(i) )/2.
14    b = sin(xAxis) + cos(xAxis)/2.
15
16    write(*,'(f8.4, 2x, f8.4, 2x, f8.4)') &
17       [ (xAxis(i), a(i), b(i), i=-N,N) ]
18 end program array_expressions1
```

Listing 2.29 src/Chapter2/array_expressions1.f90

Note that the standard does not impose a specific order in which the elements of the result array for the expression are to be created. This allows compilers to apply hardware-specific optimizations (e.g. vectorization/parallelization). For this to be possible, all array expressions are completely evaluated, before the result is assigned to any variable. This makes array expressions behave differently from do-loop constructs which superficially seem equivalent to the array expression (so one needs to carefully examine any data dependencies between the different iterations of the do-loops when translating between the two forms of syntax). This was not the case for the two array expression examples above (*lines 12* and *14* in the listing), which could have also been written equivalently with a do-loop (although we recommend the previous, compact version):

```
do i=-N,N
   a(i) = sin( xAxis(i) )
   b(i) = sin( xAxis(i) ) + cos( xAxis(i) )/2.
enddo
```

However, the expression:

```
a(-(N-1):(N-1)) = ( a(-N:(N-2)) + a(-(N-2):N)) /2.
```

which assigns to each interior element of a an average value computed using its left and right neighbours, is **not** equivalent to the loop:

```
do i=-(N-1),(N-1)
   a(i) = ( a(i-1) + a(i+1) )/2.
enddo
```

We demonstrated above that some intrinsic functions ($\boxed{\texttt{sin}}$, $\boxed{\texttt{cos}}$, etc.) accept a scalar, as well as a whole array, as their argument.[68] Such functions are known in Fortran as `elemental`, and can also be defined by the programmer, for derived types, or for specific types of arrays. We provide a brief example for this, in Sect. 3.4.

2.6.7 Using Arrays for Flow-Control

Another array-oriented feature in modern Fortran consists of two specialized flow-control constructs. Just as the `if`, `case`, and `do` were demonstrated to produce more compact code when working with scalars, for arrays the `where` and `forall` constructs can be used to simplify array expressions, and to further avoid the need for manually expanding the expressions (with loops and element-based statements). As a general note, both of these constructs can be named and nested (see Metcalf et al. [10] for details).

2.6.7.1 `where` Construct

The `where` construct can be used to restrict an array assignment only to elements which satisfy a given criterion. It is also known as *masked array assignment*. In many ways, it is the array-oriented equivalent of the `if`-construct, discussed for scalars. In its basic form, the syntax of `where` reads:

```
where ( <logicalArray> )
   array1 = <array_expression1>
   array2 = <array_expression2>
   ...
end where
```

where `logicalArray`, `array1`, `array2`, etc., must have the same shape, and `logicalArray` may also be a logical expression (for example, comparing array elements to some scalar value).

For example, assume we have two arrays a and b, and that we want to copy inside b the a-values[69] that are lower than some scalar value `threshold`. This can be easily achieved with the `where` construct, as follows:

```
program where_construct1
   implicit none
   integer, parameter :: N = 7
   character(len=100) :: outFormat
   integer :: i, j
   real :: a(N,N) = 0, b(N,N) = 0, threshold = 0.5, &
           c(N,N) = 0, d(N,N) = 0 ! used in next examples

   ! write some values in a
   call random_number( a )
```

---

[68] Programmers familiar with C++ can think of this as a restricted form of function overloading.

[69] `random_number` is an intrinsic subroutine, described in Sect. 2.7.2.

```
 ! Create dynamic format, with internal-file(=string) outFormat.
 ! This way, the format is adjusted automatically if N changes.
 write(outFormat, *)"(", N,"(x, f8.2))"

 write(*,'(a)')"a = "
 write(*, fmt=outFormat) &
       ( (a(i,j), j=1,N), i=1,N )

 ! ** Masked array-assignment **
 where( a > threshold )
      b = a
 end where

 write(*,'(/,a)')"b (after masked assignment) = "
 write(*, fmt=outFormat) ( (b(i,j), j=1,N), i=1,N )

 end program where_construct1
```

Listing 2.30 | src/Chapter2/where_construct1.f90 |

Similar to the if-construct, the where-construct could have been compacted, in this case, to a single line (since a single array assignment statement was present):

```
where( a > threshold ) b = a
```

Next, suppose we also want to copy over to array c the values of a that are smaller than half the threshold. We can extend the where-construct with an elsewhere(logicalArray) construct, similar to the elseif-branches we showed for if:

```
where( a > threshold )
   b = a
elsewhere( a < threshold/2. )
   c = a
end where
```

As a final extension of our example, let us assume that we want to copy over to array d the remaining values of a, which satisfy neither of the criteria (like the else-branch of if). This is achieved again with an elsewhere-branch, which does not have a logicalArray associated, as in:

```
where( a > threshold )
   b = a
elsewhere( a < threshold/2. )
   c = a
elsewhere
   d = a
end where
```

The logical arrays which define the masks (for the where- or elsewhere-branches) are first evaluated, and then the array assignments are performed in sequence, masked by the logical arrays (i.e. no assignment is performed for elements where the mask is | .false. |). This implies that, even if some assignments would alter the data used for evaluating the mask array,[70] such changes will not affect the remainder of the where-construct, for which the initially evaluated mask will be used.

---

[70] In our examples above, this would mean changing elements of a.

2.6.7.2 The do Concurrent Construct

The do concurrent construct (introduced in Fortran 2008) can also be used for
improving the performance and conciseness of array operations. Strictly speaking,
the construct is more general, as it can also be used to work with scalar data. However,
we discuss it here, as it is particularly useful for arrays, and also because it effectively
supersedes another array-oriented construct (forall), which we do not cover in
this text.[71]

We begin our brief discussion of this construct with a warning: as for many
Fortran 2008 features, support for do concurrent was, at the time of writing, still
incipient.[72]

The syntax of the construct is as follows:

```
do concurrent( [type_spec ::] list_of_indices_with_ranges &
   [, scalar_mask_expression] )
   statement1
   statement2
     .  .  .
end do
```

where list_of_indices_with_ranges can be an index range specification (as would appear after a normal do-loop), or a comma-separated list of
such specifications (in which case, the construct is equivalent to a set of nested
loops). We discuss the optional type_spec at the end of this section. The
scalar_mask_expression, when present, is useful for restricting the statement application only to values of indices for which the expression evaluates to
.true. . This is illustrated in the following example, where elements of matrix a
which belong to a checkerboard pattern are copied to matrix b:

```
 1  program do_concurrent_checkerboard_selection
 2     implicit none
 3     integer, parameter :: DOUBLE_REAL = selected_real_kind(15, 307)
 4     integer, parameter :: N = 5 ! side-length of the matrices
 5     integer :: i, j ! dummy-indices
 6     real(kind=DOUBLE_REAL), dimension(N,N) :: a, b ! the matrices
 7     character(len=100) :: outFormat
 8
 9     ! Create dynamic format, using internal file
10     write(outFormat, *)"(", N,"(x, f8.2))"
11     ! Initialize matrix a to some random values
12     call random_number( a )
13
14     ! Pattern-selection with do concurrent
15     do concurrent( i=1:N, j=1:N, mod(i+j, 2)==1 )
16        b(i,j) = a(i,j)
17     end do
18
19     ! Print matrix b
```

---

[71] In many ways, forall is a more restricted version of do concurrent, which is why we
prefer to describe only the latter. The syntax is very similar for both constructs. See, e.g. Metcalf
et al. [10] for more details on forall.

[72] That being said, we found that both gfortran (version 4.7.2) and ifort (version 13.0.0)
support this construct, with the exception of the type specification. Check the documentation of your compiler, for any flags that may need to be added to enable this feature (e.g.
−ftree−parallelize−loops=n , with n being the number of parallel threads

(for gfortran), or −parallel (for ifort)

```
20 |   write(*,'(/,a)')"b ="
21 |   write(*, fmt=outFormat) ( (b(i,j), j=1,N), i=1,N )
22 | end program do_concurrent_checkerboard_selection
```

Listing 2.31 `src/Chapter2/do_concurrent_checkerboard_selec-` `tion.f90`

Syntactically, the construct in *lines 15–17* in the previous listing could have been written using nested do-loops and an if, as in:

```
do i=1,N
   do j=1,N
      if( mod(i+j, 2)==1 ) then
         b(i,j) = a(i,j)
      end if
   end do
end do
```

so the version using do concurrent is obviously more compact. More importantly, the construct also enables some compiler optimizations with respect to the version using nested do-loops. There is a tradeoff, of course, because the restrictions on do concurrent do make it less general. Some of these (*restrictions*) are things that the compiler can check (and issue compile-time error if they are violated), while others cannot be checked automatically, and the programmer *guarantees* that they are satisfied.[73] For example:

- Most *restrictions* relate to preventing the programmer to branch outside the do concurrent-construct. Examples of mechanisms which can cause such branches are return, go to, exit, cycle, or err= (for error-handling). A safe rule of thumb is to avoid these statements.[74]
- Calling other procedures from the body of the construct is allowed, as long as these procedures are *pure*. This notion, discussed in more detail in the next chapter, implies that the procedure has *no side effects*; examples of side effects which would render procedures impure are:

  – altering program's state, in a global entity, or locally to the procedure, which may be used next time the procedure is called
  – producing output during one iteration, which is read during another iteration

- The programmer also *guarantees* to the compiler that there are no data dependencies between iterations (through shared variables, data allocated in one iteration and de-allocated in another iteration, or writing and reading data from an external channel in different iterations)

Given these limitations, using do concurrent may require some additional effort. However, for applications where performance is a priority, this is time

---

[73] Therefore, the program may successfully compile, but still contain bugs, if some of these implied guarantees do not actually hold!

[74] Strictly speaking, those which reference a labelled statement are allowed, as long as that statement is still within the do concurrent-construct.

well-spent, since it forces the programmer to re-structure the algorithms in ways which are favorable for parallelization at later stages (more about this in Sect. 5.3).

An interesting last note about this construct is that the standard also allows to specify the type of the indices within the construct (the type is always `integer`, but the `kind`-parameter can be customized). This is very convenient, since it brings type declarations closer to the point where the variables are used (otherwise these indices would need to be declared at the beginning of the (sub)program, as done in the previous example-program). For example, the pattern-selection portion in the earlier example-program could be written as:

```
do concurrent( integer :: l=1:N, m=1:N, mod(l+m, 2) == 1)
   b(l,m) = a(l,m)
end do
```

Note that, at the time of writing, most compilers still do not support this. However, it should be allowed in the near future.

2.6.8 Memory Allocation and Dynamic Arrays

In the examples so far, we only showed how to work with arrays whose shape is known at compile-time. This is often not the case in real applications, where this information may be the result of some computations, or may even be provided by the user at runtime. If this were a book about C++, now would definitively be the place to discuss pointers. In Fortran, however, this is not necessary[75] for dynamic-size arrays, which are supported through a simpler (and faster) mechanism, discussed in this section.

We often use the terms *static* and *dynamic* when discussing how memory is reserved for data entities. Generally speaking, memory for *static* objects is automatically managed by the OS. Examples of static entities are static global variables (defined through the `module`-facility, discussed later), variables local to a procedure, and procedure arguments (also covered later). Contrarily, *dynamic* objects require the programmer to explicitly make requests for acquiring and releasing regions of memory. Therefore, whereas for working with normal (static) arrays only a declaration is necessary, the workflow for dynamic arrays involves three steps:

1. **declaration**: Dynamic arrays are declared similarly to normal arrays. For example, a dynamic version of array `bigArray` (see Sect. 2.6.1) is given below:

```
integer, dimension(:,:,:), allocatable :: bigArray
```

---

[75] Pointers are still useful in many contexts, like for constructing more advanced data structures. They too are supported in Fortran, via the `pointer`-attribute (but Fortran pointers carry more information and restrictions than their C/C++ counterparts). We do not discuss this issue in this text—see, e.g. Metcalf et al. [10] or Chapman [1].

Note that there are two notable differences in the dynamic version:

a. the *shape* of the array is not specified; instead, only the rank is declared (encoded as the number of $\boxed{:}$-characters in the list within the parentheses)
b. the `allocatable`-attribute needs to be added, to clarify that this is a dynamic array

2. **allocation**: Before working with array elements is allowed, memory has to be allocated, so that the exact shape of the array is specified. This is done with the `allocate`-statement, which has the form:

```
allocate( list_of_objects_with_shapes [, stat=statCode] )
```

where `statCode` is an (optional) integer scalar, set to zero by the system if the allocation was successful, or to some positive value if an error occurred (such as not enough memory to hold the arrays requested), and `list_of_objects_with_shapes` is a list of arrays, each followed by the explicit shape in parentheses (as would normally appear after the `dimension`-attribute if the arrays were static). For example, the following statements allocate the dynamic versions of arrays `xArray`, `bigArray`, and `zArray`, from Sect. 2.6.1:

```
integer :: statCode
allocate( xArray(16), bigArray(520,520,520), zArray(-100:100), stat=statCode )
```

After allocation, one can work with these arrays normally, as discussed before for the static case.

3. **deallocation**: A last concern related to dynamic arrays is to release the memory to the system, as soon as it is not needed by the program anymore. This is a highly recommended practice, both for performance reasons (because it reduces the amount of bookkeeping at runtime), and for increasing the readability of the programs (to signal the fact that the data is not used in other parts of the program). This step is achieved with the `deallocate`-function, which has the syntax:

```
deallocate( list_of_objects [, stat=statCode] )
```

where `statCode` has the same error-signalling role as before, and `list_of_objects` is a list of arrays. For example, the following statement releases the memory allocated above, for the arrays `xArray`, `bigArray`, and `zArray`:

```
deallocate( xArray, bigArray, zArray, stat=statCode )
```

Note that it is an error to attempt allocating an already-allocated array, or deallocating an already-deallocated (or never allocated) array. The allocation status of an array may become difficult to track in larger programs, especially if the array is part of the global data and used by many procedures. The `allocated` intrinsic function can be used in such cases. For example:

```
allocated( xArray )
```

will return `.false.` before the `allocate`-call above, and after the `deallocate`-call; it will return `.true.`, however, between these two calls. Interestingly, since Fortran 2003, it is not necessary [13] to use this intrinsic function when we want to assign to the allocatable array another array (or array expression): in that case, allocation to the correct shape is automatically done by the Fortran runtime.

Exercise 9 (*Array transversal order and performance*) Earlier in this chapter, we mentioned that array element order dictates the optimal array-transversal order for obtaining good performance. To test this, write a program which adds two cubic $3D$-arrays (a and b), using nested do-loops. Measure the time required for the program to complete, for two different transversal scenarios:

```
do i=1,N
  do j=1,N
    do k=1,N
      a(i,j,k) = a(i,j,k) + b(i,j,k)
    enddo
  enddo
enddo

do k=1,N
  do j=1,N
    do i=1,N
      a(i,j,k) = a(i,j,k) + b(i,j,k)
    enddo
  enddo
enddo
```

Hints:

- The length of the cube's side (N) should be large enough to be representative for a real-world scenarios (i.e. the whole arrays should not fit in the cache). For example, take $N = 813$, and 32bit real array elements. It is easier to use `allocatable` arrays.[a]
- To improve the accuracy of the result, wrap the code above within another loop, so that the operations are performed, say, $N_{repetitions} = 30$ times.[b]
- It is also instructive to test the programs with several compilers, because some highly-optimizing compilers (like `ifort`) may recognize performance "bugs" like these in simple programs, and correct the problem

internally (but this can fail in more complex scenarios, so learning about these issues is still valuable). Also, compilers can simply "optimize away" code when the computation results are not used, so try to print some elements of a at the end of the computation.

---

[a]Most systems have some limits for the size of static data ("stack size"). Therefore, large static arrays would require adjusting these limits and, possibly, adjusting the "memory model" through compiler flags.

[b]This reduces the effect of system noise, and it also provides a "poor man's" solution for reducing the relative importance of the (de)allocation overhead—a more accurate approach is to benchmark the computational parts exclusively, using techniques discussed later, in Sect. 2.7.

2.7 More Intrinsic Procedures

In the course of our discussion so far, we have already mentioned some of the many intrinsic procedures offered by Fortran. In this section, we describe a few additional ones, which would not easily fit into the previous sections, but are nonetheless common practice. We discuss later (in Chap. 3) how to define custom procedures.

2.7.1 Acquiring Date and Time Information

Some ESS applications need to be concerned with the current date and time. The date_and_time intrinsic subroutine is appropriate for this. When calling this, one can pass (as an argument) an integer-array, of size 8 or more. The Fortran-runtime will then fill the components with integer-values, as described in Table 2.3.

A very common application is timing a certain portion of code, as a quick way for profiling parts of a program. In principle, using date_and_time before and after the part of the algorithm to be profiled could be used, but this limits the time

Table 2.3 Data inserted into components of curr_date_and_time

Component #	Meaning	Component #	Meaning
1	Year	5	Hour
2	Month	6	Minutes
3	Day	7	Seconds
4	Time difference (minutes) w.r.t. GMT	8	Milliseconds

resolution that can be achieved. Fortran also has the `cpu_time` intrinsic for such purposes, which provides microsecond precision on many platforms.

A complete program, demonstrating these functions, is given below:

```fortran
program working_with_date_and_time
  implicit none
  ! for date_and_time-call
  integer :: dateAndTimeArray(8)
  ! for cpu_time-call
  real :: timeStart, timeEnd
  ! variables for expensive loop
  integer :: mySum=0, i

  call date_and_time(values=dateAndTimeArray)
  print*,"dateAndTimeArray =", dateAndTimeArray

  call cpu_time(time=timeStart)
  ! expensive loop
  do i=1, 1000000000
     mySum = mySum + mySum/i
  end do
  call cpu_time(time=timeEnd)
  print*,"Time for expensive loop =", timeEnd-timeStart,"seconds",&
       ", mySum =", mySum
end program working_with_date_and_time
```

Listing 2.32 | `src/Chapter2/working_with_date_and_time.f90`

Some precautions apply to uses of `cpu_time`:

- results are generally non-portable (since the resolution is not standardized, to allow higher precision for platforms which support it)
- even if no other demanding programs seem to be running on the system, the timing results will fluctuate, due to ever-present "system noise" (the OS needs to continuously run some internal programs, to maintain proper operation)
- the function is not useful for parallel applications; for example, in a parallel program using OpenMP, the `omp_get_wtime`-subroutine should be used instead
- while convenient for quick tests, this approach to profiling does not scale (just as `print`-based debugging does not scale well for complex bugs); many manufacturers, as well as open-source projects, offer much more convenient tools for complex scenarios.

2.7.2 Random Number Generators (RNGs)

Statistical methods form the basis of many powerful algorithms in ESS. For example, stochastic parameterizations are commonly used in models, to simulate the effects of processes at smaller spatial scales (clouds, convection, etc.), which are not resolved by the (usually severely coarsened) model mesh. A basic necessity for many such algorithms is the ability to generate sequences of random numbers. This may seem a simple technicality but, in fact, it invites a philosophical question, since computer algorithms are supposed to produce deterministic outcomes.[76]

Nonetheless, many algorithms can produce sequences which are often "sufficiently random", despite being deterministic. Fortran implementations also provide such algorithms, via the `random_number` intrinsic subroutine. The following program uses it to estimate π.

```
7   program rng_estimate_pi
8     implicit none
9     integer, parameter :: NUM_DRAWS_TOTAL=1e7
10    integer :: countDrawsInCircle=0, i
11    real :: randomPosition(2)
12    integer :: seedArray(16)
13
14    ! quick method to fill seedArray
15    call date_and_time(values=seedArray(1:8))
16    call date_and_time(values=seedArray(9:16))
17    print*, seedArray
18    ! seed the RNG
19    call random_seed(put=seedArray)
20
21    do i=1, NUM_DRAWS_TOTAL
22       call random_number( randomPosition )
23       if( (randomPosition(1)**2 + randomPosition(2)**2) < 1.0 ) then
24          countDrawsInCircle = countDrawsInCircle + 1
25       end if
26    end do
27    print*,"estimated pi =", &
28          4.0*( real(countDrawsInCircle) / real(NUM_DRAWS_TOTAL))
29  end program rng_estimate_pi
```

Listing 2.33 `src/Chapter2/rng_estimate_pi.f90`

Note that we used another intrinsic subroutine (`random_seed`), to compensate for the deterministic nature of the *random number generator* (RNG).[77] To link with the previous discussion, we use two calls to the `date_and_time` intrinsic, to obtain a seed array.[78] This is not an "industrial-strength" solution, since the date information is not completely random (and some components like the time zone are in fact constant).[79]

The algorithm itself is based on placing random points within a square $2D$-domain, and checking what fraction of those fall within the largest quarter-of-circle inscribed in the square. This is a classical example of what is known as the Monte-Carlo approach to simulation.

---

[76] This is fundamentally different from randomness in the physical sense, which is driven by the quantum-probabilistic processes at the atomic scale. These effects are then amplified at the mesoscopic scales, due to the large number of degrees of freedom of the system (e.g. climate system, see Hasselmann [6]).

[77] In situations where perfect reproducibility of results is necessary, the seeding step could be skipped. However, a more scientifically-robust method to achieve this is to use a sequence of random numbers large enough that the reproducibility is achieved algorithmically.

[78] You can check how large the array needs to be for your platform, by calling the seed function like `call random_seed(size=seedSize)`, where seedSize is an `integer` scalar, inside which the result of the inquiry will be placed.

[79] A better solution for seeding may be to use the entropy pool of the OS. In Linux, you can read data from the file `/dev/random` (see, e.g. Exercise 10).

Exercise 10 (*Accurate* π) Modify the previous program, so that it reliably recovers the first 7 digits after the decimal dot of π.
Hints: you will need to ensure that the variables involved have a `kind` which is accurate enough. Also, to rule out "accidental" convergence, it is a good idea to check that the convergence criterion remains satisfied for several (say, 100) Monte-Carlo draws in a row.

As a final note on this topic, for scientific applications it is often important to ensure the RNG passes certain quality criteria—usually a batch of tests. Thus, a hierarchy of RNG-algorithms exists, relative to which the `random_number` intrinsic may not be the most suitable. For an in-depth discussion of this topic, please refer to Press et al. [12].

References

1. Chapman, S.J.: Fortran 95/2003 for Scientists and Engineers. McGraw-Hill Science/Engineering/Math, New York (2007)
2. Cormen, T.H., Leiserson, C.E., Rivest, R.L., Stein, C.: Introduction to Algorithms. MIT Press, Cambridge (2009)
3. Fraedrich, K., Jansen, H., Kirk, E., Luksch, U., Lunkeit, F.: The planet simulator: towards a user friendly model. Meteorol. Z. **14**(3), 299–304 (2005)
4. Goldberg, D.: What every computer scientist should know about floating-point arithmetic. ACM Comput. Surv. **23**(1), 5–48 (1991)
5. Hager, G., Wellein, G.: Introduction to High Performance Computing for Scientists and Engineers. CRC Press, Boca Raton (2010)
6. Hasselmann, K.: Stochastic climate models Part I. Theory Tellus **28A**(6), 473–485 (1976)
7. Kirk, E., Fraedrich, K., Lunkeit, F., Ulmen, C.: The planet simulator: a coupled system of climate modules with real time visualization. Technical Report 45(7), Linköping University (2009)
8. Marshall, J., Plumb, R.A.: Atmosphere, Ocean and Climate Dynamics: An Introductory Text, 1st edn. Academic Press, Boston (2007)
9. Mehlhorn, K., Sanders, P.: Algorithms and Data Structures: The Basic Toolbox. Springer, Berlin (2010)
10. Metcalf, M., Reid, J., Cohen, M.: Modern Fortran Explained. Oxford University Press, Oxford (2011)
11. Overton, M.L.: Numerical Computing with IEEE Floating Point Arithmetic. Society for Industrial and Applied Mathematics, Philadelphia (2001)
12. Press, W.H., Teukolsky, S.A., Vetterling, W.T., Flannery, B.P.: The art of parallel scientific computing. Numerical Recipes in Fortran 90, vol. 2, 2nd edn. Cambridge University Press, Cambridge (1996). also available as http://apps.nrbook.com/fortran/index.html
13. Reid, J.: The new features of Fortran 2003. ACM SIGPLAN Fortran Forum **26**(1), 10–33 (2007)

Chapter 3
Elements of Software Engineering

In this chapter, we introduce some software-engineering approaches which make large programs viable. We start with *structured programming* (SP), which has a strong following in the Fortran community. Next, we introduce *object-oriented programming* (OOP), a style of programming that received explicit language support only relatively recently, but which we consider worth describing even in introductory texts such as ours. The chapter closes by briefly discussing some *generic programming* (GP) aspects.

3.1 Motivation

Theoretically, all programs could be structured as a simple unit, with declarations of variables at the top, followed by executable statements (calculations, assignments, etc.), as in:

```
program monolithic_program
  implicit none
  ! specification statements (e.g. variable declarations) below
  . .
  ! program statements below
  . . .
end program monolithic_program
```

Such an organization would be perfectly acceptable for the compiler, but in practice some form of modularization is always used.[1] In this chapter, we discuss general approaches for achieving modularization, and how these map to features in Fortran.

There are many high-level reasons for modularization—for instance to keep *state* (data) and *program logic* (executable statements) manageable, as well as for reducing the overall programming effort. It is important to keep these ideas in mind, since they

---

[1] Even in the example programs presented so far we used modularization, in the form of intrinsic procedures, such as for I/O operations.

© Springer-Verlag Berlin Heidelberg 2015
D.B. Chirila and G. Lohmann, *Introduction to Modern Fortran for the Earth System Sciences*, DOI 10.1007/978-3-642-37009-0_3

often provide hints on how the modularization could be performed. Some specific reasons include:

- Modularization introduces "data partitions", so that only some parts of the program are allowed to change some variables (otherwise, all data would be "global", which is a *discouraged* practice). Such "fences" are useful, because they reduce the number of entities that the programmer needs to keep in mind at any given time. Also, they can significantly reduce the effort for parallelizing the application.
- Subtasks of a program which are general enough (like computing the norm of a vector, vorticity of a vector field, or displaying a result on the screen) can be re-used in other applications, eliminating the need to start from scratch every time. Such subprograms can then be collected into libraries, shared between many applications. This practice also increases the probability (and decreases the effort) for having the subprograms thoroughly tested, which is often difficult to do for a monolithic program.
- A more mundane aspect is that Fortran requires variable declarations to precede executable statements,[2] which in a monolithic program would force the programmer to frequently alternate between the region where variables are declared and the regions where they are used. It is clearly more convenient to have both regions fit on-screen at the same time, which is why a good practice is to aim for programs and procedures which are not too long.

Modularization of software can be approached from different points of view. In SP, which is currently the norm in Fortran, this is done with a focus on the subtasks of the program. Contrarily, in OOP, which only later received extensive language support, the focus is on types and their associated operations. Another approach for keeping the complexity of software manageable is GP, where the focus is on formulating algorithms in more general forms, so that they can be applied to the entire set of data structures which satisfy the algorithm's requirements.[3] However, we devote very little space to this approach, since Fortran does not support it extensively at this stage.

3.2 Structured Programming (SP) in Fortran

From a problem-solving point of view, SP favors the *top-down* approach where, after defining the problem to be solved, the task is divided into smaller subtasks. If necessary, subtasks themselves may need to be further divided, until the work

---

[2] There is actually a new Fortran 2008 feature (block/end block—see Metcalf et al. [8]), which enables variable declarations in other parts of the code too (e.g. local variables inside a loop). However, we do not cover this feature, as extensive compiler support was still missing at the time of this writing.

[3] For example, a sorting algorithm should work with elements of integer, real, or even user-defined types, as long as a suitable binary operator (like "less than") is defined on any two elements of the type. The ideal of GP is to write the algorithm only once, reducing duplication of code (the need to maintain a different implementation of sort for each type).

to be done forms a unit that can be easily mapped onto concrete statements in the programming language.[4] Several language features support this breaking down of tasks, which we describe in this section.

3.2.1 Subprograms and Program Units

A basic level of modularization consists of *subprograms* (also known as *procedures*).[5] These are useful for isolating a well-defined part of the program's logic. For example, in an ESS model, there might be one subprogram for initializing the model state, another for updating the model state to a new time step, and yet another for writing the model state to a file. Such subtasks, when they can be isolated, are ideal for the SP-approach.[6] Similarly, these tasks can be subdivided, e.g., the update of the model state can be subdivided into dynamics, thermodynamics, etc.

In Fortran, there are two types of subprograms: the `function` and the `subroutine`. Subprograms and program data can be organized into *program units*, which are viewed by the compiler as distinct compilation tasks. Three types of program units exist:

- *main-program*: This is what we used in most of the preceding examples. Each application needs to have *exactly one* main-program, where execution begins and (ideally) ends (we will see that subprograms are also allowed to terminate the program, when exceptional situations occur).
- *external subprograms*: These can be either `functions` or `subroutines` (discussed later). Subprograms correspond to different subtasks of the overall algorithm, and they can be invoked from the main-program, or even from other subprograms. They are structured similarly to the main-program, except that they need to expose an *interface*, describing how data is transmitted to/from the calling (sub)program.
- *modules*: These are general containers, which can encapsulate data, definitions of derived types, `namelist`-groups (see Sect. 5.2.1), `interfaces` (see Sect. 3.2.3), or module subprograms.

---

[4] The workflow is somewhat analogous to Richardson's (Richardson and Lynch [9]) energy cascade in turbulence theory (replace energy with "work still to be done", and viscous dissipation with writing code).

[5] We use the terms *subprogram* and *procedure* interchangeably in this text.

[6] Strictly speaking, the term *SP* also covers the use of flow-control constructs (`if`s, loops, etc.), not only of subprograms. However, in this text we reserve the term for referring to the aspect of subprogram-based program design, since the use of flow-control constructs (instead of `goto`-statements) is nowadays taken for granted.

Note on code organization
Concerning actual code layout, it is considered good practice to arrange each program-unit in a separate file. However, this requires some knowledge about build systems, which we only cover later, in Sect. 5.1. To avoid complicating our discussion with such aspects, we organize most of the examples in this chapter in single files, which is also legal (for example, a module and a main-program using it may be placed in the same file, as long as the module appears first). However, keep in mind that this approach can become inconvenient in larger programs (imagine a single file with 1,000,000 lines of code!).

Internal subprograms: Prior to our discussion of subprograms, it is worth mentioning that main-programs, external subprograms, as well as module subprograms can contain nested *internal subprograms*. For these, the contains-keyword needs to be added, on a line of its own, after the last executable statement of the host (sub)program. Then, the internal subprogram can be added between this line and the end program(/function/subroutine)-line of the host. We do not discuss internal subprograms in more detail, however, since they are not so often used, for good reasons.[7]

Dataflow: From a high-level perspective, the execution of an application designed according to the SP-paradigm consists of a series of calls to subprograms. Dataflow mechanisms need to exist, such that the data created by one subprogram can be read (and perhaps altered) by the other subprograms. Two mechanisms for accomplishing this in Fortran are:

1. passing data through the procedure interface (i.e. arguments and, for functions, also the returned data). This is the clearer method, recommended most of the times.
2. using modules to share data encapsulated within them across procedures. This approach can also be useful sometimes, but we generally recommend considering the first approach first.[8]

In the next sections, we focus on the first mechanism (for the second one, see Sect. 3.2.7).

---

[7] In particular, internal subprograms have direct access to the data (and other internal subprograms) of their host. This form of unstructured data access may be tempting on the short-term, but generally has a negative effect on the readability of the software. Also, internal subprograms are fundamentally tied to their host, which makes them difficult to re-use in other (sub)programs (since the internal subprograms would need to be converted into normal subprograms first; however, this process may be nontrivial, especially if the above form of data access was (ab)used by the programmer).

[8] Packaging a lot of data inside modules can increase the probability of bugs, such as accidentally modifying the data from procedures that should not modify it (in contrast, passing data through the procedure interface allows more control on allowed operations). Also, subprograms which rely on much module data are generally more difficult to understand, and cannot be easily re-used.

3.2.2 *Procedures in Fortran (`function` and `subroutine`)*

As mentioned previously, two types of procedures (subprograms) exist in Fortran. A `function` takes (zero or more) arguments, and returns *exactly one* result. Function calls can simply appear as parts of more complicated expressions, wherever a value of the returned type is legal. In contrast, a `subroutine` can only be invoked (used) with an explicit `call`-statement, on a separate line. Several criteria for deciding between the two types are given in Sect. 3.2.5. For now, it is enough to know that a `function` preferably does not modify its arguments, while a `subroutine` routinely does (which also allows the latter to return more than one entity). Argument and result type for procedures may be scalar, string, or array (of intrinsic or derived type[9]). Interestingly, arguments may also be other subprograms, which enables library subprograms to call user-written subprograms.

3.2.2.1 Declaring Procedures

To illustrate the use of procedures, we provide in the next listings two possible ways of printing the prime numbers up to some values (in the code, 100). The program $\boxed{\texttt{primes\_with\_func1a}}$ (in the Listing 3.1) uses the function `isPrimeFunc1a`, which takes an integer argument, and returns a logical value, depending on whether the number is prime or not (the actual algorithm for testing whether a number is prime is not important in this context). Similarly, the program $\boxed{\texttt{primes\_with\_sub}}$ (in the Listing 3.2) uses the subroutine `isPrimeSub`, for the same effect.

```
1   logical function isPrimeFunc1a( nr )
2      implicit none
3      ! data-declarations (for interface)
4      integer, intent(in) :: nr
5      ! data-declarations (local variables)
6      integer :: i, squareRoot
7
8      if( nr <= 1 ) then
9         isPrimeFunc1a = .false.; return
10     elseif( nr == 2 ) then
11        isPrimeFunc1a = .true.; return
12     elseif( mod(nr, 2) == 0) then
13        isPrimeFunc1a = .false.; return
14     else
15        squareRoot = int( sqrt(real(nr)) )
16        do i=3, squareRoot, 2
17           if( mod(nr, i) == 0 ) then
18              isPrimeFunc1a = .false.; return
19           endif
20        enddo
21     endif
22     isPrimeFunc1a = .true.
23  end function isPrimeFunc1a
24
25  program primes_with_func1a
26     implicit none
27     integer, parameter :: N_MAX=100
28     integer :: n
29     ! declaration for function
30     ! NOT the recommended approach
31     logical isPrimeFunc1a
```

---

[9] *Derived Data Types* (DTs), alternatively named "abstract" types, are discussed in Sect. 3.3.2.

```
32
33     do n=2, N_MAX
34         if(isPrimeFunc1a(n)) print*, n
35     enddo
36 end program primes_with_func1a
```

`src/Chapter3/primes_with_func1a.f90`

```
1  subroutine isPrimeSub( nr, isPrime )
2     implicit none
3     ! data-declarations (for interface)
4     integer, intent(in) :: nr
5     logical, intent(out) :: isPrime
6     ! data-declarations (local variables)
7     integer :: i, squareRoot
8
9     if( nr <= 1 ) then
10        isPrime = .false.; return
11    elseif( nr == 2 ) then
12        isPrime = .true.; return
13    elseif( mod(nr, 2) == 0) then
14        isPrime = .false.; return
15    else
16        squareRoot = int( sqrt(real(nr)) )
17        do i=3, squareRoot, 2
18            if( mod(nr, i) == 0 ) then
19                isPrime = .false.; return
20            endif
21        enddo
22    endif
23    isPrime = .true.
24 end subroutine isPrimeSub
25
26 program primes_with_sub
27    implicit none
28    integer, parameter :: N_MAX=100
29    integer :: n
30    logical :: stat
31
32    do n=2, N_MAX
33        call isPrimeSub(n, stat)
34        if(stat) print*, n
35    enddo
36 end program primes_with_sub
```

Listing 3.2 `src/Chapter3/primes_with_sub.f90`

Let us analyze the new constructs. First of all, both `isPrimeFunc1a` and `isPrimeSub` appear as blocks above the corresponding main-programs. On *line 1/Listing 3.1* (respectively *line 1/Listing 3.2*), we have the function(subroutine)-statement (corresponding to the *function header* in C/C++ terminology). The `implicit none` statements have exactly the same role as in a main-program (to prevent the language from implicitly associating variables with data types); as for main-programs, it is recommended to use this statements at the beginning of all procedures (as well as `modules`—discussed later).

On *line 4/Listing 3.1* (respectively *4-5/Listing 3.2*), several variables are declared. These are not normal variable declarations, however. Rather, these lines define parts of the function (subroutine) interfaces—a fact marked by the use of the `intent`-attribute. Possible choices for this attribute are:

- `in` when a value only needs to be *read* within the procedure (as is the case in `isPrimeFunc1a`)
- `out` when a value is overwritten by the procedure, without being accessed beforehand (as is the case for the returned value in `isPrimeSub`)
- `inout` when the value needs to be both *read* and *written* by the procedure (not used in the examples above)

Such variables, which appear in the list of arguments of the procedure and have the intent-attribute,[10] are known as *dummy arguments*. In essence, such arguments are placeholders, waiting to be replaced by *actual arguments*, when the procedures are invoked.

In Listing 3.1, there is also an *implicit* variable declaration, which completes the interface of the function. This corresponds to the value returned by the function to the calling program. By default, this value has the same name as the function (in our example isPrimeFunc1a, of logical type—as specified within the function statement, on *line 1*).

There are actually two other equivalent methods for defining a function, which may be encountered in practice:

1. The first one, which makes the declaration explicit, is:

```
1   function isPrimeFunc1b( nr )
2     implicit none
3     ! data-declarations (for interface)
4     integer, intent(in)   :: nr
5     logical :: isPrimeFunc1b ! NOTE: return-type defined here; no 'intent'
6     !          allowed (it is effectively "out")
7     ! data-declarations (local variables)
8     integer :: i, squareRoot
```

Listing 3.3 src/Chapter3/primes_with_func1b.f90 (excerpt)

Note that, unlike the first variant, the return type of the function is specified in a separate line within the function body (*line 5*). This should *not* have an intent-attribute (since that is effectively set to out by the language).

2. The second alternative also changes the name of the function's result, to make this distinct from the function's name:

```
1   function isPrimeFunc1c( nr )   result(primStat)
2     implicit none
3     ! data-declarations (for interface)
4     integer, intent(in) :: nr
5     logical :: primStat ! NOTE: return-type declared here
6     ! data-declarations (local variables)
7     integer :: i, squareRoot
```

Listing 3.4 src/Chapter3/primes_with_func1c.f90 (excerpt)

Here, the return type of the function is also specified separately and without an intent-attribute (*line 5*). In addition, however, we also use the result-keyword (*line 1*), to change the name of the result to something different from the function's name. This can be useful when the function name is long and inconvenient to use in expressions. Also, it is *mandatory* when writing recursive functions (a topic not discussed in this book).

Returning to Listings 3.1 and 3.2, notice that two more variables are declared (i and squareRoot). Since they are introduced within the scope of the procedures

---

[10] Although some compilers may still accept procedure-argument declarations without the intent-attribute, it is always recommended to specify these attributes, to make it clear how each of the arguments is supposed to be used (good documentation); additionally, the compiler can then detect some frequent mistakes (such as accidentally overwriting a variable that is only supposed to be read).

(thus only being accessible while the procedures are executing), these are called *local variables*. We will discuss several issues related to such variables in Sect. 3.2.4.

Finally, the executable parts of the procedures (*lines 8–22* in Listing 3.1, respectively *lines 9–23* in Listing 3.2) can contain assignments, flow-control constructs, calls to other procedures, etc., similarly to the executable portions of the example programs presented previously. There are, however, two notable differences:

1. First of all, propagation of information outside the procedures is achieved by assigning the desired result value to the variables `isPrimeFunc1a` and `isPrime`, respectively.
2. Second, note the `return`-statements, which cause the procedures to stop, and execution to continue in the calling (sub)program. This is especially useful for skipping part of the procedure's code, for example when the result can be determined early on. In our case, for example, if `nr < 1`, it does not make sense to perform any additional tests, since such numbers are, by definition, not prime.[11]

3.2.2.2 Using (calling) Procedures

Assuming a procedure (which we wrote, or obtained from other sources, such as external libraries) is made known to the (sub)program where it is needed (caller), we can invoke it. This happens on *line 34* of Listing 3.1 and *line 33* of Listing 3.2, respectively. As illustrated in the examples, the calling syntax is different for functions and subroutines: while function calls can be part of expressions (whereby the function result is substituted, after which the evaluation of the expression continues as normal), subroutine calls need to appear on a *separate line*, and be preceded by the `call`-keyword. The call will result in changes (visible to the caller), for any variables passed to the subroutine which matched dummy arguments with `intent(out)` or `intent(inout)` within the subroutine definition. After the call, the new values of such variables can be used normally within the caller.

3.2.3 Procedure Interfaces

We mentioned above that a (non-intrinsic) procedure needs to be made known to the caller. This is the role of the procedure *interface*. In Listing 3.1, this was done in a "quick-and-dirty" way (*line 31*), with a type-declaration statement for the function name (indicating that the result of the function is `logical`); for the subroutine example (Listing 3.2), this issue was ignored altogether.

Such approaches, however, are not recommended in complex applications. To understand why, one needs to consider that the actual assembly code for passing

---

[11] In this case, using `return` improves performance for large values of N_MAX. However, for some simple procedures there might also be a performance penalty, as having multiple exit points from a procedure may prevent some compiler optimizations, e.g., auto-vectorization.

execution between caller and callee is usually generated at link-time. However, checking that the procedure was correctly invoked (with the right types of arguments and in the correct order) is usually the responsibility of the compiler. For the compiler to be able to perform this task, it needs knowledge about both the call site and the declarations in the procedure (so that actual arguments can be matched against dummy arguments). Depending on how much of this information is actually available to the compiler, the interface is said to be *implicit* or *explicit*:

- `implicit` When only the types at the call site are available to the compiler (and this much is always known when the program unit of the caller is parsed), the interface is said to be *implicit*. Compilation can succeed in this situation (with only a type-declaration for functions, and "as is" for subroutines). However, relying on *implicit* interfaces is *dangerous*, since many useful compiler checks are thus effectively turned off by the programmer. To illustrate how easily this may lead to bugs, replace *line 34* (Listing 3.1) with:

```
!                        bug
         if(isPrimeFunc1a(n*1.0)) print*, n
```

Listing 3.5 | `src/Chapter3/primes_with_func_bug.f90` | (excerpt)

Note the multiplication by 1.0, which leads to a result of type `real`.[12] The program still compiles[13]; however, when executed, it does not report any prime numbers anymore. The reason can be found by analyzing (e.g. with a simple `write`-statement) what data the function `isPrimeFunc1a` actually receives. On our platform, the first 3 numbers were:

```
isPrimeFunc1a: got nr = 1073741824
isPrimeFunc1a: got nr = 1077936128
isPrimeFunc1a: got nr = 1082130432
```

(instead of the expected 2, 3, 4). The perplexing numbers occur because the compiler tries to interpret a `real`-number as an `integer`.

- `explicit` When the compiler has access to both the types at the call site, as well as to the correct types of the procedure arguments, the interface is said to be *explicit*. Proper checking of interfaces can then take place, so that bugs of the type discussed above (and others) are easily detected automatically. Clearly, this is a desirable situation. It can be achieved through three main mechanisms:

1. Compilers usually interpret each program unit as a whole, so the interface will automatically be explicit, without any additional programmer effort, when *both the caller and the callee are within the same program unit.*

---

[12] Granted, this change is a little artificial in this context, but attempting to call procedures with the wrong type can happen often in complex and long-lived projects.

[13] *Some* compilers may issue a warning in our specific example, since the function and the main-program are still in the same file. If they are separated, however, this would not occur.

Specifically, this rule applies when a subprogram in a module calls another subprogram from the same module, or when a (sub)program calls one of its internal subprograms.

2. When a (sub)program includes a module with a `use`-statement, the interface is also explicit for any subprograms found in that module (we discuss this mechanism in more detail in Sect. 3.2.7).

3. If none of the previous two scenarios holds, the programmer needs to explicitly define an `interface`-block, to make the interface of an external subprogram explicit. Such a block has the form:

```
interface
    ! interface body here . . .
end interface
```

To give a specific example, the bug we introduced above (related to the `isPrimeFunc1a`-function) will be caught by the compiler if we make the interface explicit with such an `interface`-block, as in:

```
1  program primes_with_func_bug_avoided
2      implicit none
3      integer, parameter :: N_MAX=100
4      integer :: n
5
6      ! If we make the interface explicit (e.g. with the
7      ! interface-block below), the bug is easily identified.
8      interface
9          logical function isPrimeFunc1a( nr )
10             integer, intent(in) :: nr
11         end function isPrimeFunc1a
12     end interface
13
14     do n=2, N_MAX
15         !                   bug
16         if(isPrimeFunc1a(n*1.0)) print*, n
17     end do
18 end program primes_with_func_bug_avoided
```

Listing 3.6 `src/Chapter3/primes_with_func_bug_avoided.f90`
(excerpt)

The `interface`-block should contain the beginning marker (header line) of the subprogram (*line 9* in the example above), the types of the arguments along with their `intent` (*line 10*), and the end marker of the subprogram (*line 11*). Executable statements, or internal subprograms (if any) *should not* appear.

Note that the `interface`-block takes a significant part of our program. In more complex scenarios, where multiple external subprograms are used, it is not convenient to insert the interfaces within the caller, as we just did—a more usable approach, instead, is to package the interfaces within a module, and to `use` that in the caller (as discussed later, in Sect. 3.2.7).

We will discuss other uses of `interface`-blocks later: in Sect. 3.3 we show how these can support important OOP-concepts, and in Sect. 5.4 we employ them again for enabling Fortran programs to call procedures defined in C.

Exercise 11 (*Procedures for unit-conversions*) Write a function which converts degrees to radians, and a subroutine which converts radians to degrees. Test the results for a small set of input-angles (e.g. $0°$, $30°$, $45°$, $60°$ and $90°$). Does "pipelining" the function and the subroutine give back the same degree-value with which you started? If not, explain why.

3.2.3.1 Arrays and Strings as Procedure Arguments

Since most applications in ESS operate on arrays, it is important to know how these can be passed as procedure arguments. In modern Fortran, there are two recommended approaches for achieving this: *explicit-shape* and *assumed-shape* arrays. Both approaches allow the modern features of Fortran arrays, which we mentioned in Chap. 2, to be used (array sections, array expressions, etc.).[14]

Explicit-shape dummy arrays In this case, the programmer needs to explicitly pass the bounds for each dimension of the used array(s) to the procedure, via additional procedure arguments. This is shown in the following example, which takes as an argument a $2D$-array of temperature values (in °C), measured at numSites stations, and computes the overall average temperature:

```
1   real function calcAvgTempV1( inArray, startTime, endTime, numSites )
2       implicit none
3       integer, intent(in) :: startTime, endTime, numSites
4       ! explicit-shape dummy array
5       real, dimension(startTime:endTime, numSites), intent(in) :: inArray ! Celsius
6
7       calcAvgTempV1 = sum( inArray, mask=(inArray > -273.15) ) / &
8           count( mask=(inArray > -273.15) )
9   end function calcAvgTempV1
```

Listing 3.7 | `src/Chapter3/function_explicit_shape_array.f90` | (excerpt)

Note the arguments startTime, endTime, and numSites, which determine the shape for the dummy-array declaration inArray within the function (thus the shape is said to be *explicit*).

Assumed-shape dummy arrays With assumed-shape dummy arrays, there is no need for the programmer to explicitly pass the shape of the arrays through the procedure's list of arguments. Instead, the dimensions of the dummy array are declared with colons, and when the procedure is called the dummy array assumes (takes) its shape from that of the actual argument. For example, the previous function can be simplified by making inArray an assumed-shape array, as follows:

```
8    real function calcAvgTempV2 ( inArray )
9        implicit none
10       ! assumed-shape dummy array
11       real, dimension(:,:), intent(in) :: inArray ! Celsius
12
13       calcAvgTempV2 = sum( inArray, mask=(inArray > -273.15) ) / &
```

---

[14] There is also a third approach (*assumed-size* arrays), which is however *strongly discouraged* (and not covered in this text), since it provides little information about the array to the compiler, effectively disabling those high-level array features—see for example Chapman [3] if working with legacy code that uses this feature.

```
14          count( mask=(inArray > -273.15) )
15    end function calcAvgTempV2
```

Listing 3.8 | `src/Chapter3/function_assumed_shape_array.f90`
(excerpt)

One fact to keep in mind when using assumed-shape dummy arrays is that the *extents* of the actual array along each dimension are passed to the procedure, but *not the lower and upper bounds*. For example, if we called the function above as in:

```
real :: functionResult, sampleData(100:365, 20)
! ... write data into sampleData array ...
functionResult = calcAvgTempV2( sampleData ) ! call
```

the dummy array `inArray` (within the function) will assume the bounds (`1:266`, `1:20`). This was not a problem for our function, where the result is not influenced by this "shifting" of the bounds. However, in applications where this is important, the loss of "metadata" can be prevented by specifying a lower bound in the declaration for the dummy array; this bound may either be a constant (when there is a natural choice for this), or another argument. We illustrate the second approach below:

```
17    real function calcAvgTempV3( inArray, startTime )
18        implicit none
19        ! explicit lower-bound, to preserve array-shape
20        integer, intent(in) :: startTime
21        ! assumed-shape dummy array, with explicit lower-bound
22        real, dimension(startTime:, :), intent(in) :: inArray ! Celsius
23
24        calcAvgTempV3 = sum( inArray, mask=(inArray > -273.15) ) / &
25            count( mask=(inArray > -273.15) )
26    end function calcAvgTempV3
```

Listing 3.9 | `src/Chapter3/function_assumed_shape_array.f90`
(excerpt)

Character strings as procedure arguments It may sometimes also be useful to pass character strings to procedures. In such cases, it is not necessary to explicitly specify the length of the strings, since Fortran supports *assumed-length character strings*, where the `len` type parameter is set to $*$. This is illustrated in the next listing, which partly re-uses the logic from Listing 2.23, to compute the number of vowels in a string:

```
1    integer function countVowels( strng )
2        implicit none
3        character(len=*), intent(in) :: strng
4        integer :: numVowels, i
5
6        numVowels = 0 ! reset counter
7        ! it is allowed to inquire the length of the actual-argument with 'len'
8        do i=1, len(strng)
9            select case( strng(i:i) )
10           case( 'a', 'e', 'i', 'o', 'u', &
11               'A', 'E', 'I', 'O', 'U' )
12               numVowels = numVowels + 1
13           end select
14       end do
15       countVowels = numVowels
16   end function countVowels
```

Listing 3.10 | `src/Chapter3/function_assumed_length_string.f90` (excerpt)

3.2.3.2 Argument Keywords

Although we would usually try to keep the number of arguments in procedures small, this may not always be possible (for example, when subprograms from a library are used). In such situations, it is all too easy to make the mistake of passing arguments in the wrong order (especially if adjacent dummy arguments in the procedure prototype have the same type—in which case the compiler would not catch the semantic error). A very useful Fortran feature for avoiding such problems is that *the names given to the dummy arguments can actually be used as* keywords (tags). With this technique, the order in which arguments are specified at the call site is not important. To give an example, the following subroutine samples the function $z(x, y) = \cos(x^2 + xy)e^{-0.05(x^2+y^2)}$, with a resolution `res`, along a rectangular plane section $(x, y) \in [-5.0, 5.0] \times [-10.0, 10.0]$:

```fortran
subroutine sampleFunctionToFileV1( xMin, xMax, yMin, yMax, res, outFileName )
   implicit none
   real, intent(in) :: xMin, xMax, yMin, yMax
   integer, intent(in) :: res
   character(len=*), intent(in) :: outFileName
   integer :: i, j, outFileID
   real :: x, y, a, b, c, d

   ! ensure 'res' received a valid value (should be >=2)
   if( res < 2 ) then
      write(*,'(a,1x,i0,1x,a)') "Error: res =", res, "is invalid! Aborting."; stop
   end if
   open(newunit=outFileID, file=outFileName, status="replace")
   ! evaluate scaling-coefficients
   a = (xMax-xMin)/(res-1); b = (res*xMin-xMax)/(res-1)
   c = (yMax-yMin)/(res-1); d = (res*yMin-yMax)/(res-1)

   do i=1, res
      do j=1, res
         x = a*i+b; y = c*j+d ! scale to real
         write(outFileID, '(3(f16.8))') x, y, cos( x*(x+y) )*exp( -0.05*(x**2+y**2) )
      end do
      write(outFileID,*) ! newline for GnuPlot
   end do
   close(outFileID)
end subroutine sampleFunctionToFileV1
```

Listing 3.11 `src/Chapter3/sample_surface.f90` (excerpt)

Using keywords, the following call still produces the intended result,[15] even if the order of the arguments does not coincide with that in the function header:

```fortran
call sampleFunctionToFileV1( outFileName="test_func_sample.dat", &
     xMin=-5., yMin=-5., xMax=10., yMax=10., res=200 )
```

Listing 3.12 `src/Chapter3/sample_surface.f90` (excerpt)

In addition, the use of keywords also serves as good documentation (provided the author of the procedure used meaningful names for the dummy arguments). We also demonstrated using keywords in the earlier discussion of file-based I/O.

3.2.3.3 Optional Arguments

Another method for making work with custom procedures easier is to make (some of) the arguments optional. This makes sense when sensible default values can be

---

[15] The resulting data file can be easily visualized, for example, in (gnuplot), using the command:

```
splot 'sampling_test1.dat' using 1:2:3 with pm3d
```

chosen for some arguments, which are appropriate most of the time, but we still want
to allow advanced users to tune the values. In Fortran, the corresponding dummy
arguments need to be declared with the additional `optional`-attribute. Then, within
the executable part of the procedure, it is possible to check (with the `present`
intrinsic function) if the optional argument was actually specified at the call site or
not.

To provide an example, let us re-write the previous procedure, so that some default
values are chosen for the `res` and `outFileName` arguments, when they are not
specified at the call site:

```fortran
 1  subroutine sampleFunctionToFileV2( xMin, xMax, yMin, yMax, res, outFileName )
 2    implicit none
 3    real, intent(in) :: xMin, xMax, yMin, yMax
 4    integer, optional, intent(in) :: res
 5    character(len=*), optional, intent(in) :: outFileName
 6    integer :: i, j, outFileID
 7    real :: x, y, a, b, c, d
 8    ! default values for optional arguments
 9    integer, parameter :: defaultRes = 300
10    character(len=*), parameter :: defaultOutFileName = 'test_func_sample.dat'
11    ! local vars for optional-args
12    integer :: actualRes
13    character(len=256) :: actualOutFileName ! need to specify length
14
15    ! initialize local vars corresponding to optional-args. If the caller
16    ! actually provided values for these args, we copy them; otherwise, we use
17    ! the default values...
18    ! ...res
19    if( present(res) ) then
20        actualRes = res
21    else
22        actualRes = defaultRes
23    endif
24    ! ...outFileName
25    if( present(outFileName) ) then
26        actualOutFileName = outFileName
27    else
28        actualOutFileName = defaultOutFileName
29    endif
30
31    ! ensure 'actualRes' value is valid (should be >=2)
32    if( actualRes < 2 ) then
33        write(*,'(a,1x,i0,1x,a)') "Error: res =", res, "is invalid! Aborting."; stop
34    end if
35
36    ! open output-file
37    open(newunit=outFileID, file=trim(adjustl(actualOutFileName)), status="replace")
38    ! evaluate scaling-coefficients
39    a=(xMax-xMin)/(actualRes-1); b=(actualRes*xMin-xMax)/(actualRes-1)
40    c=(yMax-yMin)/(actualRes-1); d=(actualRes*yMin-yMax)/(actualRes-1)
41
42    do i=1, actualRes
43        do j=1, actualRes
44            x = a*i+b; y = c*j+d ! scale to real
45            write(outFileID, '(3(f16.8))') x, y, cos( x*(x+y) )*exp( -0.05*(x**2+y**2) )
46        end do
47        write(outFileID,*) ! newline for GnuPlot
48    end do
49    close(outFileID)
50  end subroutine sampleFunctionToFileV2
```

Listing 3.13 `src/Chapter3/sample_surface_optional_args.f90` (excerpt)

With this new version of the subroutine, the following call becomes valid:

```fortran
! call which does not specify the res-argument
call sampleFunctionToFileV2( xMin=-5., xMax=5., yMin=-10., yMax=10., &
    outFileName='test_func_sample_lowres.dat' )
```

Listing 3.14 `src/Chapter3/sample_surface_optional_args.f90` (excerpt)

In fact, we already illustrated previously another excellent use for optional arguments, when we discussed error-handling for some of the intrinsic procedures. This
is implemented with the optional `stat` argument (which can also be used as a keyword), which the subroutine updates, to mark if any error condition occurred during
its execution. This is preferable to the alternative mechanism (whereby the subroutine simply causes the program to crash if an error condition occurred) since it allows

the caller to take control over the error-recovery process (perhaps there is a method to recover from the error or, if not, maybe some operations are necessary, such as saving data generated up to that point, closing files, etc.). Hence, the use of optional arguments for this type of error-handling is considered good practice, especially for libraries to be used by other programmers. A complete discussion is outside the scope of this book—see, e.g., Chapman [3] for details.

3.2.3.4 Passing Other Procedures as Procedure-Arguments

The attentive reader may have noticed a practical issue with our function-sampling example so far: it can only sample a *specific* function, which is hard-coded. Clearly, with the exception of the single line where the expression for the function to be sampled actually appears, the rest of the subroutine is generic enough to apply to any function which takes two `real`-values and returns another `real` (mathematically— $f : \mathbb{R}^2 \to \mathbb{R}$). It would be tedious (and a major source of code duplication) if we had to write a different version of the subroutine for each function of two real variables we want to sample. Luckily, Fortran allows other procedures to be passed as arguments to other functions, via a mechanism that works similarly to the C/C++ *function pointers* or the C++ *function objects*. The next version of our function-sampling[16] subroutine illustrates this:

```fortran
subroutine sampleFunctionToFileV3 ( xMin, xMax, yMin, yMax, func, outFileName )
  implicit none
  real, intent(in) :: xMin, xMax, yMin, yMax
  ! IFACE for procedure-argument
  interface
     real function func( x, y )
        real, intent(in) :: x, y
     end function func
  end interface
  character(len=*), intent(in) :: outFileName
  integer :: i, j, outFileID
  real :: x, y, a, b, c, d
  integer :: res = 300

  open(newunit=outFileID, file=outFileName, status="replace")
  ! evaluate scaling-coefficients
  a=(xMax-xMin)/(res-1); b=(res*xMin-xMax)/(res-1)
  c=(yMax-yMin)/(res-1); d=(res*yMin-yMax)/(res-1)

  do i=1, res
     do j=1, res
        x = a*i+b; y = c*j+d ! scale to real
        write(outFileID, '(3(f16.8))') x, y, func( x, y )
     end do
     write(outFileID,*)
  end do
  close(outFileID)
end subroutine sampleFunctionToFileV3
```

Listing 3.15 `src/Chapter3/sample_any_surface.f90` (excerpt)

Note that, for such uses, an `interface`-block takes the place of the usual declarations of arguments. For testing, we provide some functions (`test_func1` and `test_func2`), in a module[17]:

---

[16] The resolution was hard-coded here, for brevity. However, the reader can find a more complete implementation, which re-introduces adjustable resolution and also demonstrates error-handling, in the program `sample_any_surface_with_error_recovery.f90`, in the source code repository.

[17] For convenience, we anticipate the discussion of `modules`—we cover these shortly, in Sect. 3.2.7.

```
1    module TestFunctions2D
2    contains
3      real function evalFunc1( x, y )
4        real, intent(in) :: x, y
5        evalFunc1 = cos( x*(x+y) )*exp( -0.05*(x**2+y**2) )
6      end function evalFunc1
7
8      real function evalFunc2( x, y )
9        real, intent(in) :: x, y
10       evalFunc2 = cos( x+y )
11     end function evalFunc2
12   end module TestFunctions2D
```

Listing 3.16 `src/Chapter3/sample_any_surface.f90` (excerpt)

Finally, the sampling subroutine may be called as in:

```
1    program sample_any_surface
2      use TestFunctions2D
3      implicit none
4      interface
5        subroutine sampleFunctionToFileV3( xMin, xMax, yMin, yMax, func, &
6          outFileName )
7          real, intent(in) :: xMin, xMax, yMin, yMax
8          interface
9            real function func( x, y )
10             real, intent(in) :: x, y
11           end function func
12         end interface
13         character(len=*), intent(in) :: outFileName
14       end subroutine sampleFunctionToFileV3
15     end interface
16
17     ! sample function 1
18     call sampleFunctionToFileV3( xMin=-5., xMax=5., yMin=-10., yMax=10., &
19       func=evalFunc1, outFileName="sampling_func1.dat" )
20
21     ! sample function 2
22     call sampleFunctionToFileV3( xMin=-5., xMax=5., yMin=-10., yMax=10., &
23       func=evalFunc2, outFileName="sampling_func2.dat" )
24   end program sample_any_surface
```

Listing 3.17 `src/Chapter3/sample_any_surface.f90` (excerpt)

For readers unfamiliar with these techniques, it is interesting to pause and think about the succession of procedure calls taking place: program `sample_any_surface` calls the subroutine `sampleFunctionToFileV3` which, in turn, calls whatever function was passed to it as an argument by the user (above—`evalFunc1` and `evalFunc2`). In practice, procedures such as `sampleFunctionToFileV3` are often part of libraries for some domain-specific problem (function integration, minimization, etc.). Passing procedures as arguments is then essential for library authors, who cannot anticipate all the functions which library users may wish to combine with the library.

Note on performance

Some of the most useful places where calls to procedure arguments could be made are within (nested) loops. In these cases, using this technique as illustrated above can degrade performance (especially when individual invocations of the user-supplied function are relatively inexpensive). The reason is that, to maximize efficiency in this case (inexpensive functions), the compiler should be able to simply copy the body of the user-supplied function inside the code generated for the caller (a process named *function inlining*). When this optimization occurs, explicit function calls are avoided, which can significantly improve performance, without affecting the correctness of the algorithms. The increased performance is due to the fact that explicit procedure calls carry a non-zero overhead of their own; if the actual "work" (computations) inside the functions is so small that it is comparable to this overhead, a failure of the compiler to perform inlining can severely hurt performance. Just as passing function pointers in C/C++, passing procedures as arguments in Fortran can prevent function inlining (although this also depends on the quality of the compiler).

Readers for which these issues become relevant should consult the documentation of their compiler (most compilers can be configured to provide feedback about optimizations which failed). However, such work is generally best left to the later stages of development (algorithm correctness should always be the top priority). Also, if optimization is indeed necessary, it should be guided by profiling (see, e.g., Hager and Wellein [7]).

`stop`-**statement** Throughout several versions of the function-sampling example presented so far, we showed how to abort the program in case a fatal error occurs within the subroutine, using the `stop`-statement. This is different from `return`, as it *terminates the entire execution of the program*, no matter where the `stop`-statement is encountered (i.e. in the main-program or a procedure). While it is useful to be aware of this mechanism note that, by using it, the caller is no longer offered the chance to recover from the error, which is *not a recommended practice*. Due to limited space, we do not demonstrate the recommended procedure in the text—please refer to program `sample_any_surface_with_error_recovery.f90` in the source code repository for a sample implementation.

In case there are several places where `stop`-statements are used, it is possible to add a constant character string, or a number (with maximum 5 digits), to help clarify precisely which statement occurrence caused the program to abort (useful for debugging).

3.2.4 Procedure-Local Data

We already mentioned that it is possible to declare variables local to a procedure, which are not part of the interface. These are normally temporary variables, used for storing intermediate data values, that play a role in the computation/operations taking place within the procedure. For example, in Listings 3.1 and 3.2, variables `i` and `squareRoot` were local. The issues of *data scope* and *persistence* across subsequent procedure calls are important in this context, and are explained in this section. In addition, we also present a mechanism for more convenient allocation of variable-size arrays within procedures—the *automatic arrays*.

3.2.4.1 Data Scope

Data entities have an attached scope, consisting of the places (in the code of the application) where the data can be accessed and—for variables—modified. Restricting the scope of entities is useful, since it allows the programmer to minimize unwanted interactions between different program units (therefore making the code more maintainable). In general, entities defined within a program unit (main-program, external procedure, or module), are not accessible from other program units. This rule, however, does not prevent access *within a given program unit*. For example, a module procedure can access all the data entities within the same module (as discussed in more detail in Sect. 3.2.7). Also, *internal* procedures can access data of their host, as well as call other internal procedures of the same host ("host association").

3.2.4.2 Persistence of Data and the ⎡save⎤ Attribute

A peculiar characteristic of Fortran, which may catch programmers familiar with other languages off-guard, relates to the state of variables across subsequent invocations of procedures. For example, in C/C++ any changes to local variables during one call will normally not influence the values of the variables in subsequent calls to the same function.[18] If we want the function to save these changes internally, the variable declaration needs to be preceded by the static keyword.

The equivalent keyword in Fortran is the save type attribute. *However*, variables will also gain this attribute implicitly, if they are initialized in the subroutine. Due to this, it is very easy to accidentally get saved variables, in situations where this is not intended (which is most times), for example for local variables used to store intermediate state within the procedure. We demonstrate this in the following listing, where the local variable tmpSum is initialized this way (the function is supposed to simply sum all the elements in the received array, which is assumed to be one-dimensional):

```
1    real function sumArrayElementsV1( inArray )
2        implicit none
3        real, dimension(:), intent(in) :: inArray
4        real :: tmpSum=0.0
5        integer :: i
6
7        do i=1, size(inArray)
8            tmpSum = tmpSum + inArray(i)
9        end do
10       sumArrayElementsV1 = tmpSum ! collect result
11   end function sumArrayElementsV1
12
13   program implicitly_saved_var_buggy
14       implicit none
15       interface
16           real function sumArrayElementsV1( inArray )
17               real, dimension(:), intent(in) :: inArray
18           end function sumArrayElementsV1
19       end interface
20       real, dimension(4) :: testArray = [ 1.0, 3.0, 2.1, 7.9 ]
21
22       write(*,*)"sum1 =", sumArrayElementsV1( testArray )
```

---

[18] In C/C++, there is no separation of procedures into functions and subroutines— only the former are allowed, although the void return type allows emulation of Fortran subroutines. We will discuss in more detail this classification in Sect. 3.2.5.

```
23    write(*,*) "sum2 =", sumArrayElementsV1( testArray )
24  end program implicitly_saved_var_buggy
```

Listing 3.18 `src/Chapter3/implicitly_saved_var_buggy.f90`

The function `sumArrayElementsV1` will return a different result the second time it is called, because the initial value of `tmpSum` will not be zero anymore. The obvious solution to such problems is to initialize procedure-local variables via separate statements, in the executable section of the procedure's code (see file `implicitly_saved_var_fixed.f90`, in the source code repository).

A more pertinent use of `saved` variables is to keep some internal counters, for debugging purposes. For example, the local variable `evenCountDebug` in the following subroutine keeps track of the number of invocations for which an even number was passed as input:

```
1   subroutine compute4thPower( inNumber, outNumber )
2     implicit none
3     integer, intent(in) :: inNumber
4     integer, intent(out) :: outNumber
5     integer, save :: evenCountDebug=0
6     if( mod(inNumber, 2) == 0 ) then
7       evenCountDebug = evenCountDebug + 1
8     end if
9     write(*,'(a,i0,/)') "@ compute4thPower: evenCountDebug =", evenCountDebug
10
11    ! code for subroutine's functionality (not much in this example)
12    outNumber = inNumber**4
13  end subroutine compute4thPower
```

Listing 3.19 `src/Chapter3/save_attribute_debugging_counters.f90`
(excerpt)

However, it is a good idea not to rely on the `save`-attribute in released code, since it can make programs more difficult to understand. Also, it can make future parallelization work difficult, since assumptions about the order of procedure calls (which may not hold in parallel environments) are too easy to incorporate.

3.2.4.3 Automatic Arrays

We discussed previously, in Sect. 2.6.8, how to reserve memory for arrays whose shape is only known at runtime. Such arrays are very useful for storing the large data structures in applications, which may need to be read and modified by many different procedures. A pitfall of such arrays, however, is that they require explicit memory-management effort from the programmer, who should remember to allocate the arrays before use, and de-allocate them as soon as they are no longer needed.

In the context of procedures, however, *automatic arrays* provide a convenient, procedure-*local* alternative to allocatable arrays. They are especially suited for creating arrays whose shape also becomes known only at runtime, but which are *not needed outside the procedure*. With this restriction, the programmer can simply declare these as normal arrays (except that the bounds of the dimensions are variables); the Fortran runtime system then takes care of the lower-level (de)allocation details in the background. For example, the following function uses the automatic

array `workArray`, to implement the *counting sort* algorithm, that takes an array of
n integers $\in [0, k]$, and returns another array with the sorted integers[19]:

```fortran
function countingSort( inArray, k )
   implicit none
   integer, dimension(:), intent(in) :: inArray
   integer, intent(in) :: k ! max possible element-value in 'inArray'
   integer, dimension( size(inArray) ) :: countingSort ! func result
   ! NOTE: automatic array (shape depends on function's arguments)
   integer :: workArray(0:k)
   integer :: i

   ! Automatic arrays cannot be initialized on declaration-line
   workArray = 0
   ! Place histogram of input inside workArray.
   do i=1, size(inArray)
      workArray( inArray(i) ) = workArray( inArray(i) ) + 1
   end do
   ! Accumulate in workArray(i) the # of elements less than or
   ! equal to i.
   do i=1, k
      workArray(i) = workArray(i) + workArray(i-1)
   end do
   ! Place elements at appropriate position in output-array.
   do i=size( inArray ), 1, -1
      countingSort( workArray(inArray(i)) ) = inArray(i)
      workArray( inArray(i) ) = workArray( inArray(i) ) - 1
   end do
end function countingSort
```

Listing 3.20 | `src/Chapter3/function_with_automatic_array.f90` | (excerpt)

There are several restrictions on the use of automatic objects:

- they may not be initialized in a type-declaration statement (which is why we also initialize `workArray` in the executable section of the function in the previous example)
- they may not have the `save`-attribute, i.e., an automatic array cannot persist across multiple calls to the same procedure; indeed, because the shape of the array would probably be different for different calls to the procedure, the very concept of persistence does not make sense here
- they may not be used in `namelist`-groups (a topic we discuss later, in Sect. 5.2.1).

3.2.5 *Function* or *Subroutine?*

The presence of two categories of subprograms in Fortran may be confusing for programmers familiar with other languages. For example, in C/C++, only functions are supported, but the returned entity can also have the special `void` type, when there is no value to return. Fortran has no such keyword, which is a first reason for distinguishing two subprogram categories.

For the practicing programmer, there are several criteria which help in deciding between the two categories.

Number of entities to be returned When there is only one entity to be returned, it is natural to write a `function`. Otherwise (when returning 0 or > 1 objects) a `subroutine` should be used.

---

[19] For our purpose here, the details of the algorithm are not important (but we refer the interested readers to Cormen et al. [6] for more details).

Convenience of calling Another criterion is how conveniently the procedure can be invoked: a `function` can simply appear as part of an expression, while a `subroutine` needs to be `called` on a separate line, and only afterwards can the variables modified be used in an expression.

Programming convention related to side effects The two categories of procedures can also be used to enforce (manually, via coding conventions) a very useful classification with respect to side effects, which can significantly improve the readability of the software. Specifically, a `function` would have no side effects. We mentioned in passing, in the previous chapter, what side effects are. Below is a more complete list of operations which qualify as side effects (and are *generally to be avoided if possible*, for reasons we will discuss shortly):

- *modification of procedure arguments*: In general, procedures are much easier to analyze and debug if they do not modify their arguments (behaving like functions in mathematics).
- *modification of data external to the procedure*: For example, if a `module` is used, which contains public data declarations, modifying such data would qualify as a side effect.[20] A related type of side effect is when an internal procedure modifies data entities of its host.[21]
- *saved variables (explicit/implicit)*: We discussed earlier in this chapter how variables can gain the `save`-attribute. The problem with this is that it becomes too easy to make subsequent invocations of the procedure depend on each other (which would be a problem for parallelization, for example).
- *including `stop`-statements*: As discussed above, such statements allow a procedure to directly terminate the entire application, which may be fine for a serial program, but raises severe problems in the parallel case, where it could lead to data corruption.
- *performing I/O on units external to the procedure*: Most usual units are external, so the only channel on which I/O can be performed without introducing a side effect is an internal file, declared locally to the procedure.
- *calling other procedures with side effects*: Of course, this also applies to calling procedures which are passed as arguments, as discussed earlier. A further restriction is that called procedures are not allowed to modify variables which are declared as read-only in the current (caller) procedure.

It is clear from the list above that there are quite a few requirements to keep track of before deciding that a procedure does not have side effects. Nonetheless, the additional effort that the programmer needs to invest for keeping this property usually pays off through easier debugging when a problem appears. Also, to make the task of checking for side effects easier, the `pure`-keyword was introduced in Fortran, which causes the compiler to check if the side effects mentioned are indeed not present. The keyword needs to be added in the header line, as in:

---

[20] The procedure is said to gain access to the module's data through *use association* in this case.

[21] Data access is through what is known as *host association* then.

Table 3.1 Criteria for deciding whether to write a `function` or a `subroutine`

		# of returned arguments	
		1	$\neq 1$
Can be written without side effects	Y	Pure function	Pure subroutine
	N	Impure subroutine	Impure subroutine

Note that "impure" subroutines are simply normal subroutines (declared without the `pure`-keyword)

```
pure <resT> function doFoo( <argLst> )
   ! ... function body ...
   ! resT: type of result
   ! argLst: list of arguments
end function doFoo
```

Listing 3.21 Declaring a pure function

```
pure subroutine doBar( <argLst> )
   ! ... subroutine body ...
   ! argLst: list of arguments
end subroutine doBar
```

Listing 3.22 Declaring a pure subroutine

We provide, in file `forbidden_side_effects_for_pure.f90` (in the source code repository) a series of intentional violations of side-effect-related rules, to allow the reader to test the specific error messages reported by the specific compiler installed.

We provide in Table 3.1 a system based on two criteria, for deciding what type of procedure to write for any given task. In particular, we emphasize the use of `pure`-procedures if possible, which make the programs easier to understand, and also make things easier for parallelization. Another application of `pure` for array operations is discussed in Sect. 3.4 (`elemental` procedures).

3.2.6 Avoiding Name Clashes for Procedures

Given the large number of intrinsic procedures in Fortran,[22] and that large applications usually consist of many custom-written procedures themselves, name clashes are a real concern (whereby a programmer defines a procedure with the same name as one of the intrinsics). The behavior in such cases (i.e. which version is selected by the compiler) can become a source of confusion, which we attempt to summarily clarify here.[23]

If the intention of the programmer is to select the custom procedure instead of the intrinsic one, it is enough to make the interface of the custom procedure *explicit*, as

---

[22] In addition, compiler vendors are allowed to provide additional intrinsic subroutines, not specified by the language standard.

[23] Of course, a good rule of thumb for avoiding this issue altogether is to avoid such name clashes altogether, unless there is a very good justification (such as when extending the intrinsic procedure, to allow working with derived types in the same way as with the standard types—see Sect. 3.3.5).

was described in Sect. 3.2.3. It is also possible to select the custom procedure when the interface is *implicit*, by adding a external procedure1, procedure2, ... specification statement. Note, however, that since this does not add any information about the interface, the compiler still cannot perform proper checking of argument (therefore this method is *not recommended*).

If, however, the intention of the programmer is to select the intrinsic procedure, this can be achieved with the intrinsic-statement. This needs to appear before the executable statements in the (sub)program, as in:

```
! make it clear that we refer to the intrinsic procedure
! (assuming a name clash with a custom procedure could occur)
intrinsic min ! list of intrinsic procedures
```

Although sometimes useful in handling name clashes, this practice is not routinely used for specifying all the intrinsic procedures used in the application, which would be too tedious. However, a very good use of this feature is for simplifying the porting of applications to a different platform, *when the application uses vendor-dependent intrinsic functions for some well-defined reason*. A different compiler, which does not have the non-standard extension, would then issue a more useful error message.

3.2.7 Modules

Another type of program unit, which is very useful for structuring nontrivial applications better, is the module. In particular, these can be used to group items related to a particular task (which is also why most Fortran software libraries are conveniently exposed as a module[24]). Items which can be packaged in a module are:

- **global data**: constants needed across multiple (sub)programs; variables may also need to be shared for efficiency sometimes, although in general they are best avoided;
- **subprograms**: this practice which has the additional benefit of making the interface of the subprogram *explicit* automatically, leading to shorter programs due to elimination of interface-blocks—for example, the program sample_any_ surface and many others from previous sections of this chapter;
- interface-**blocks**: these are relevant for *external* procedures, not defined inline within the module, and also for many OOP techniques, discussed in Sect. 3.3;
- namelist-**groups**: this use is covered in Sect. 5.2.1;
- **derived types** and **corresponding operations**: these are essential in OOP—see Sect. 3.3.2.

A module can contain a *specification part* in the beginning, although this is optional. After that, it can include (also optionally) a *procedures part* (in which case it is necessary to include the contains-keyword on a separate line, to mark that what follows are procedure definitions). This structure is shown below:

---

[24] The library implementation may actually use a hierarchy of modules internally, but often a single module needs to be presented to the user.

```
module ModuleName
  !'use'-statements, to include other modules (optional)
  implicit none
  ! specification statements, for example:
  ! * global variables/constants,
  ! * interface blocks,
  ! * namelist groups, or
  ! * declarations for derived types
contains
  ! procedure definitions (functions or subroutines)
end module ModuleName
```

Note the `implicit none` line, which has the same role as discussed earlier (enforcing explicit type declarations for variables).[25]

Entities declared within a `module` can be made available to (sub)programs (or even another `module`), with the `use`-statement, as in:

```
use ModuleName [, only : moduleEntity1, moduleEntity2, ... ]
```

where the portion inside brackets indicates that it is possible to select *individual entities* from a module (otherwise, all `public`[26] entities of the module will become available). Such a restriction is useful when only a few entities are needed from a large module, or to document the source of specific entities for developers of the program/module,[27] when several modules are used. The `use`-statement, when present, should be the first to appear in the body of the subroutine or module (even before the `implicit none` statement).

It is also possible to create a *local alias* when including a specific entity from a module, to improve clarity or to avoid name clashes. This is also done in the `use`-statement, as shown below:

```
use ModuleName [, only : localAlias1 => moduleEntity1, ... ]
```

As a first `module`-example, let us package in a more convenient form the code for obtaining portable precision for the `real` type (first discussed in Sect. 2.3.4):

```
10  module RealKinds
11    implicit none
12    ! KIND-parameters for real-values
13    integer, parameter :: &
14        R_SP = selected_real_kind( 6, 37 ), &
15        R_DP = selected_real_kind( 15, 307 ), &
16        R_QP = selected_real_kind( 33, 4931 )
17    ! Edit-descriptors for real-values
18    character(len=*), parameter :: R_SP_FMT ="f0.6", &
19        R_DP_FMT = "f0.15", R_QP_FMT = "f0.33"
20
21  contains
22    ! Module-subprogram.
23    subroutine printSupportedRealKinds()
24      write(*,'(a)') "** START: printSupportedRealKinds **"
25      if( R_SP > 0 ) then
26        write(*,'(a,i0,a)') "single-prec. supported (kind=", R_SP, ")"
27      else
28        write(*,'(a,i0,a)') "single-prec. MISSING! (kind=", R_SP, ")"
29      end if
30
```

---

[25] Interestingly, by adding such a line at the beginning of the module, it is not necessary to include it inside the procedure declarations (if there are any)—although it also does not hurt to keep that habit.

[26] Access control for modules will be discussed shortly.

[27] Who would be spared the effort to read through all of the `used` module to find a specific data/procedure definition.

```
31        if( R_DP > 0 ) then
32            write(*,'(a,i0,a)') "double-prec. supported (kind=", R_DP, ")"
33        else
34            write(*,'(a,i0,a)') "double-prec. MISSING!  (kind=", R_DP, ")"
35        end if
36
37        if( R_QP > 0 ) then
38            write(*,'(a,i0,a)') "quad-prec.   supported (kind=", R_QP, ")"
39        else
40            write(*,'(a,i0,a)') "quad-prec.   MISSING!  (kind=", R_QP, ")"
41        end if
42        write(*,'(a)') "** END: printSupportedRealKinds **"
43    end subroutine printSupportedRealKinds
44 end module RealKinds
```

Listing 3.23 | `src/Chapter3/portable_real_kinds.f90` | (excerpt)

The module can then be used in (sub)programs, as shown in the next listing.
Note that, because of the `only` keyword, the constants R_SP and R_SP_FMT will
not be available to the program. Also, a module-procedure alias is defined, such that
`printSupportedRealKinds` (defined in the `module`) can be used under the
name `showFloatingPointDiagnostics`.

```
46 program portable_real_kinds
47     use RealKinds, only : R_DP, R_QP, &
48         R_DP_FMT, R_QP_FMT, &
49         showFloatingPoingDiagnostics => printSupportedRealKinds
50     implicit none
51
52     real(R_DP) a
53     real(R_QP) b
54
55     a = sqrt(2.0_R_DP); b = sqrt(2.0_R_QP)
56
57     call showFloatingPoingDiagnostics()
58
59     write(*,'(a,1x,' // R_DP_FMT // ')') "sqrt(2) in    double-precision is", a
60     write(*,'(a,1x,' // R_QP_FMT // ')') "sqrt(2) in quadruple-precision is", b
61 end program portable_real_kinds
```

Listing 3.24 | `src/Chapter3/portable_real_kinds.f90` | (excerpt)

Note that when the module and the (sub)program/other module which uses it
are in the same file (as is the case in our example), most compilers require the
module to appear before the point where it is actually used. Packaging both entities
in the same file is, however, *not recommended* (except for small tests). Indeed, in
the present application the RealKinds-module can only be used by (sub)programs
and other modules in the same file, which is clearly too restrictive. We present a
better approach later, while covering build systems such as *GNU Make* (gmake)
(see Sect. 5.1). However, for conciseness we use mostly the "single-file" approach
throughout this chapter.

Persistence of data within a module As of Fortran 2008, variables declared within
a module implicitly have the `save`-attribute (so it is not necessary to include
this keyword on the declaration line, as was required by previous iterations of the
standard).

3.2.7.1 Access-Control for Module Entities

A fundamental principle in software engineering is *information hiding*. Although this
has negative connotations in a world which emphasizes transparency, it is actually *a
service for the library users* in programming. The idea is to hide the implementation

details of the library from the programs which use it, so that access to the functionality only proceeds through a well-defined interface[28] (and *not*, for example, by reading and/or writing directly internal data structures of the library). This restriction is beneficial, since library developers are free to improve the library through re-structuring the internal implementation, and the users of the library do not need to modify their programs (assuming the interface was kept invariant). Also, if there are more libraries with the same interface, the users can also painlessly switch to another library (a good example is *Basic Linear Algebra Subprograms* (BLAS)—see Sect. 5.6.2).

Information hiding is well supported in Fortran via `modules`, where one can specify an access-control attribute, to set the visibility of the entity *from program units where the **module** is **used***. When an entity is visible to another program unit, it is furthermore possible to restrict its uses to *read-only*, with the additional attribute `protected`. Thus, in order of increasing rights, entities can be:

- `private` : not visible outside the module. This should be used for any data and procedures only relevant to the module implementation (but irrelevant to the users).
- `public, protected` [29]: visible outside the module, but *only as read-only*. This is useful for exposing things like internal counters of the `module`, which may be needed by users, but should not be modified by them. Note that `protected` only complements `public`, so both are necessary (unless the latter attribute is gained through a `module`-wide statement, as described below).
- `public` : entity is visible outside the module, and *can be both read and written* in the program units which `use` the `module`. This is necessary for procedures relevant to the users, but one should seek to minimize the number of variables with this attribute, to enforce information hiding.

All types of entities which can be packaged in a `module` can be augmented with such access specifiers (except `interface`-blocks, which are `public`). By default, if no access-control is used, the entities are given the `public`-status, so everything is visible to the outside code. To change the default policy to `private` for a specific module, the `private`-keyword can be included (on a line of its own), in the *specification part* of the module. Finally, these attributes can also be used in the form of statements (appearing in the specification part of the module); this can increase readability, by listing the public interface in a single place. Usage of these access-control features is demonstrated below:

```
9    module TestModule
10      implicit none
11      private  ! Change to restrictive default-access.
12      integer, public, protected :: countA=0, countB=0
13      integer                    :: countC=0
14
15      public executeTaskA, executeTaskB ! Specify public-interface of the module.
16    contains
```

---

[28] In this context, "interface" represents the entire set of library procedures that can be called by the program.

[29] C++ programmers should note that "protected" here is not the same notion as in that language, where the keyword relates to visibility in relation to inheritance.

```
17  subroutine executeTaskA()
18    call executeTaskC()
19    countA = countA + 1 ! increment debug counter
20  end subroutine executeTaskA
21
22  subroutine executeTaskB()
23    countB = countB + 1 ! increment debug counter
24  end subroutine executeTaskB
25
26  subroutine executeTaskC()
27    countC = countC + 1 ! increment debug counter
28  end subroutine executeTaskC
29 end module TestModule
```

Listing 3.25 | `src/Chapter3/test_access_control_in_modules.f90` | (excerpt)

There, we defined three subroutines (one private to the module), and some variables (`countA`, `countB` and `countC`) to keep track of the number of invocations of these subroutines (for debugging). The `module` can then be used by programs, for example:

```
31 program test_access_control_in_modules
32   use TestModule
33   implicit none
34
35   call executeTaskA() ! Some calls
36   call executeTaskB() ! to
37   call executeTaskA() ! module-subroutines.
38   ! Compilation-error if enabled (subroutine not visible, because it is made
39   ! 'private' in the module)
40   !call executeTaskC()
41
42   ! Display debugging-counters.
43   write(*,'(a,1x,i0,1x,a)') '"executeTaskA" was called', countA, 'times'
44   write(*,'(a,1x,i0,1x,a)') '"executeTaskB" was called', countB, 'times'
45   ! Compilation-error if enabled (module-variable not visible, because it is
46   ! made 'private' in the module)
47   !write(*,'(a,1x,i0,1x,a)') '"executeTaskC" was called', countC, 'times'
48 end program test_access_control_in_modules
```

Listing 3.26 | `src/Chapter3/test_access_control_in_modules.f90` | (excerpt)

3.3 Elements of Object-Oriented Programming (OOP)

OOP is an alternative to the SP methodology, which became widespread during the last decades. In this section, we provide a concise introduction to the fundamental concepts of OOP (*encapsulation*, *inheritance*, and *polymorphism*), and illustrate how these can be implemented in Fortran 2003. For more information, see Rouson et al. [10] (for an advanced discussion on the subject of scientific software engineering, in Fortran and C++), Booch et al. [2] (overview of OOP-concepts), or Clerman and Spector [5] (collection of related practices and recommendations, as applied to Fortran).

DISCLAIMER
Many of the features described in this section require a compiler with (at least partial) support for Fortran 2003. To ensure that you have the appropriate compiler see, for example, Chivers and Sleightholme [4] (or consult the documentation of your compiler for up-to-date information).

3.3.1 Solution Process with OOP

While SP favors a *top-down* approach, in OOP the *bottom-up* view is emphasized. Specifically, one does not begin by identifying the subtasks that need to be executed, but rather by identifying the entities involved in the problem (*vector*, *matrix*, etc.), and the messages that need to be supported by each entity (for a *matrix—transpose*, *compute eigenvalues*, *multiply-by-vector*, etc.). The messages are supported, in practice, via procedures associated to the entity ("methods" in C++). The process of grouping data with procedures, also known as *encapsulation*, is fundamental to OOP. The idea is to create self-contained entities which handle their own internal state, increasing the potential for software reuse.

Following the bottom-up approach, the entities created can be combined into more complex entities, using *aggregation* ("has a"-relation), or *inheritance* ("is a"-relation). The goal is to create a whole "ecosystem" of entity types, in terms of which the final solution to our target problem can be expressed more naturally. Taking, for example, an implicit numerical model in ESS (where the numerical method dictates the use of matrix inversions), we may start by identifying *vector* as a basic entity, then use aggregation to define the *matrix*-entity as a collection of vectors. Provided that the appropriate methods are implemented, the solution to the problem can be written more concisely using these new abstractions[30] than could be done while discussing in terms of `integer`, `real`, or `character` variables and arrays.

This different approach to building software that OOP proposes does come with a relatively steep learning curve, especially for programmers already comfortable with SP.[31] Obviously, much operational software (including most current ESS models) was created using the SP-approach. Therefore, it is reasonable to ask, at the onset of our discussion, *what justifies this learning effort?*

The issues that OOP aims to address relate to *scaling* of software projects, which are becoming increasingly complex. ESS is an excellent example where this phenomenon is taking place, as there is a continuous push towards more realistic, highly-coupled models. The increase in model realism unfortunately also increases the complexity of the code, mainly because there are more variables that need to be considered, which lead to intricate inter-connections between remote parts of the code. This situation tends to become harder to manage in SP. OOP can help divide this data space into entities which are easier to understand and maintain. Information hiding, which we already mentioned (Sect. 3.2.7), should be used between entities, so that they can influence each other only via a well-defined interface (and *not* by directly overwriting another entity's data). Such "fences" allow the implementation of each entity to be improved (without forcing changes upon other dependent entities), or

---

[30] Note that this example is meant only for illustrative purposes—for linear algebra there is already a wealth of good software available (see Sect. 5.6.2).

[31] This does not mean that knowing OOP in one language guarantees a smooth transition—unfortunately, there are no strict one-to-one mappings of terminology from traditional OOP languages (like C++ or Java) to equivalent Fortran constructs, hence confusion can occur.

even to replace large parts of the applications more easily (in ESS—to use a different ocean model for example).[32]

Such partitions are intuitive, so OOP is not a revolution in software engineering, but rather an evolutionary step, which allows for more powerful management of abstractions.[33] The Fortran language-standardization committee acknowledged these developments, by including many features of OOP into the modern revisions (especially Fortran 2003).

3.3.2 Derived Data Types (DTs)

In modeling problems, we often encounter entities which are more complex than what can be described by a variable or array of homogeneous, intrinsic type. Fortran accommodates such situations by allowing user-defined types (also known as *Derived Data Types* (DTs) or *abstract data types* (ADTs)). These provide the means to package entities of different types (scalars, arrays, other DTs, etc.) into a single logical unit; they are the closest correspondent to traditional OOP *classes*,[34] and provide the basic vehicle for *encapsulation*.

3.3.2.1 Defining DTs

DT-definitions are marked by a pair of $\boxed{\texttt{type DtName}}$ $\boxed{\texttt{end type DtName}}$ statements (on distinct lines). Between those, we have declarations for any data entities from which the DT is composed of, as well as declarations for type-bound procedures. To illustrate, we can define the following DT to represent $2D$-vectors:

```
5   ! DT-definitions usually included in modules.
6   module Vec2D_class
7     implicit none
8
9     type Vec2D ! Below: declarations for data-members
```

---

[32] All this assumes that the interfaces between the modules are invariant.

[33] Such evolution phenomena reflect the attempts of the software community to keep up with the large leaps in the capabilities of the underlying hardware and in user expectations. Assembly language was an evolution from machine opcodes, which took place when the hardware became too complex to manage directly in terms of opcodes. Similarly, high-level SP-languages appeared as a second evolutionary step, when assembly was not sufficient anymore to handle the software- and requirements-complexity. Nowadays, we have OOP, but functional programming is also gaining more ground.

[34] Pioneering efforts in OOP using Fortran (e.g. Akin [1]) used `modules` to emulate classes, since they can encapsulate both data and procedures. However, since there is no concept of multiple instances of a `module`, only "class-wide" data is supported (corresponding to `static` class members in C++). With these tools, it was still possible to emulate "usual" classes by making the static data an array, which held all "instances" of the class. However, this condemned the programmer of the module to handle tedious memory management for that array; since there is a more convenient alternative in Fortran 2003, we do not describe this practice in details, and instead view `modules` largely as C++ namespaces.

```
10        real :: mU = 0., mV = 0.
11     contains     ! Below: declarations for type-bound procedures
12        procedure :: getMagnitude => getMagnitudeVec2D
13     end type Vec2D
14
15  contains
16     real function getMagnitudeVec2D( this )
17        class(Vec2D), intent(in) :: this
18        getMagnitudeVec2D = sqrt( this%mU**2 + this%mV**2 )
19     end function getMagnitudeVec2D
20  end module Vec2D_class
```

Listing 3.27 | `src/Chapter3/dt_basic_demo.f90` | (excerpt)

To encourage their re-use, each DT-definition is usually placed in a `module` (ideally one module for each DT, which is why some authors, including us here, customarily add the suffix `_class`).

In many ways, a DTs resembles a `module`. First, there is a *specification part* (only *line 10* in the example above), where the data components are specified. In our case, each instance of `Vec2D` will have two `real`-variables. Assuming `myVec` is a variable of type `Vec2D`, we can access the components as in: `myVec%mU` and `myVec%mV`.[35]

Second, separated by a `contains`-statement, we have the optional *procedures part* (*line 12* above). This is however not exactly the same as for a `module`, since *only declarations appear*—the actual code for the procedures is elsewhere. Support for such procedures (named *type-bound procedures* or *methods*) was introduced in Fortran 2003. The interface for them needs to be *explicit* (so they can be either module procedures, or external procedures with an `interface`-block). Our initial version of `Vec2D` has a single method—the function `getMagnitude`, which is in fact an *alias* to the function `getMagnitudeVec2D`, at *lines 16–19* in the host module. That looks like a normal function definition, except the dummy argument (`this`) is declared differently: in the position where we used intrinsic types until now, we need to use `class(<DtName>)` (*line 17*), to tell the compiler that we refer to a derived type. This argument, named *passed-object dummy argument*, will correspond to the object for which the method is called, when it is bound to the DT. The binding of this dummy argument is triggered by *line 12* in our example. The general syntax for such bindings is:

```
procedure [(interfaceName)] [ListOfBindAttrs ::] bindName [=> procedureName]
```

where:

- `interfaceName`,[36] if specified, can be used to implement the Fortran equivalent of *abstract base classes*.[37] However, this is a topic outside our short tutorial here (see, e.g., Clerman and Spector [5] for details).

---

[35] Of course, this is allowed only if the data is `public`. This is the default policy, which we leverage here for brevity (we will soon discuss alternatives more consistent with the information hiding principle).

[36] When this argument is present, it needs to be surrounded by round brackets.

[37] These are special DTs, which are relevant in *inheritance*-hierarchies, for fixating the interface for DTs in such a hierarchy, but deferring the actual implementation of methods to the leaf-DTs.

- `ListOfBindAttrs`, if specified, is a comma-separated list of attributes, where the elements may be `public` or `private` (related to information hiding, as we will discuss), `pass` or `nopass` (related to the *passed-object dummy argument*), or `non_overridable` (related to *inheritance*).
- `bindName`, *the only mandatory argument*, represents the name under which the procedure will be available in the code using our DT. When `procedureName` is absent, `bindName` needs to correspond to the name of a procedure which is actually implemented (otherwise, it can be any name).
- `procedureName`, when specified, represents the name of an actual procedure (to which `bindName` will be an alias, for the code using the DT).

Returning to `this`, it is important to know that, by default, it does not appear on the invocation line, as it is silently added by the compiler. By default, it corresponds to the first argument in definition of the actual procedure (as was the case in our example). However, it is possible to fine-tune this process, with the binding attribute-list mentioned above:

- When the `nopass` binding attribute is used, the object is not passed to the procedure anymore. This is not so common, but may be useful as an optimization, for the case when the method does not actually need access to the data of the object (or there is no such data).
- By using the `pass(dummyArgName)` binding attribute, it is possible to select an argument other than the first, to be forwarded to the procedure as a *passed-object dummy argument*. Obviously, `dummyArgName` needs to be replaced by the real name of a dummy argument in the procedure. A situation where this technique is useful is *operator overloading* (Sect. 3.3.4).

Finally, note that any name can be chosen for the *passed-object dummy argument*. However, a common convention is to name it $\boxed{\texttt{this}}$, to match the syntax of other OOP-languages.

3.3.2.2 Instantiating and Using DTs

The `module` hosting the DT can be used by programs, to declare variables and constants of the new type, as in:

```
22  program test_driver_a
23    use Vec2D_class
24    implicit none
25
26    type(Vec2D) :: A ! Implicit initialization
27    type(Vec2D) :: B = Vec2D(mU=1.1, mV=9.4) ! can use mU&mV as keywords
28    type(Vec2D), parameter :: C = Vec2D(1.0, 3.2)
29
30    ! Accessing components of a data-type.
31    write(*, '(3(a,1x,f0.3))') &
32      "A%U =", A%mU, ", A%V =", A%mV, ", A%magnitude =", A%getMagnitude(), &
33      "B%U =", B%mU, ", B%V =", B%mV, ", B%magnitude =", B%getMagnitude(), &
34      "C%U =", C%mU, ", C%V =", C%mV, ", C%magnitude =", C%getMagnitude()
35  end program test_driver_a
```

Listing 3.28 `src/Chapter3/dt_basic_demo.f90` (excerpt)

For declarations (*lines 26–28*), the type of data is specified with `type` (`<DtName>`)-constructs (instead of, e.g., `integer`). Regarding initialization, it is possible to initialize directly on the declaration line (B)—as usual, this is required for constants (C). However, note that we did not explicitly initialize A. This is to demonstrate a mechanism which is available for DTs but not for implicit types—*default values*. Whereas there is no standard method for assigning a conventional default value (e.g. 0) to variables of intrinsic types, for DTs we can specify such values (mU = mV = 0—see *line 10* in Listing 3.27, where the DT was defined).

To make DT-initializations possible, Fortran provides *implicit constructors* behind-the-scenes. These look like function calls, where the names of the data members of the DT can be used as keywords, to improve readability (*line 27* above). It is possible to write custom constructors, if the default ones are not sufficient (but with some important observations, discussed below).

Similarly, the analogue of *destructors* in other languages are `final`-procedures. These should be written when pointers are used, or when special actions are necessary when the DT ceases to exist. The finalizers are also specified after the `contains`-statement in the DT definition (although, strictly speaking, they are not type-bound procedures). The syntax for them is:

```
final :: ListOfProcedures
```

We give an example for a case when such procedures are useful later (Sect. 5.2.2), while discussing `netCDF`-output.

Methods of the DT can be invoked[38] in a similar way as one would reference a data member, with the name of the object, followed by $\boxed{\%}$, and then by the name of the method with arguments in brackets. For example, in *lines 32–34* of Listing 3.28, we call the `getMagnitude` method of `Vec2D`. We can apply here the previous discussion on *passed-object dummy arguments*: although no arguments seem to be specified to the method in this example, we know from the definition of the method that there should be one argument—`this` is silently added by the compiler (receiving A, B, and C as an actual argument—see *lines 32, 33,* and *34* respectively).

3.3.2.3 Access-Control and Information Hiding

For demonstration purposes, we left the internal data of the type `Vec2D` above accessible from the main-program. However, in doing so we violated the *data hiding* and *encapsulation* principles of OOP, which undermines many of the benefits of the paradigm: for example, if the maintainers of the `Vec2D` DT decide that a representation in polar coordinates (r, θ) would be more efficient than (x, y), they cannot simply make this change without considering that all users would also need to modify their programs.[39] To remedy this problem, it is best to fine-tune exactly what is visible

---

[38] In OOP jargon, method invocations are also referred to as "sending a message" to the object.

[39] For our simplified example, this would not be a big problem. However, in large projects, where the DT is used by many developers, such disruptive changes can cause significant friction.

to other program units, with judicious use of the `private` and `public` keywords. For our example, we could adopt the following DT-definition:

```
1   ! NOTE: This DT declaration is too restrictive (see following discussion).
2   type, public :: Vec2D ! DT explicitly declared "public"
3     private   ! Make internal data "private" by default.
4     real :: mU = 0., mV = 0.
5   contains
6     private ! Make methods "private" by default.
7             ! (good practice for the case when we have
8             ! implementation-specific methods, that the user
9             ! does not need to know about).
10    procedure, public :: getMagnitude
11  end type Vec2D
```

Listing 3.29 Declaration for Vec2D, using more restrictive access-control

Note that we added a `private`-statement to both sections of the DT-definition (data and methods[40] have independent access-control), to change the default policy. The restriction of access, however, does bring a small cost: now we have the responsibility of designing proper mechanisms for interacting with the DT.

First, note that with the type definition above it is not possible to construct a vector with some custom components. In fact, there is no way for code in other program units to create non-zero vectors, which makes our implementation not very useful! The core of the problem is that the compiler is not allowed to provide an implicit constructor, because the data members are now `private`.

1. A first possible solution to this issue is to define a custom constructor. The mechanics for doing this are different in Fortran compared to other languages, as *user-defined constructors are not type-bound procedures.*[41] Instead, the binding of the constructor to the type is achieved by a named `interface`-block (also known as "generic interface" in Fortran), with the name of the DT we wish to construct.

2. Alternatively, we can declare a normal type-bound procedure (named, for example, `init`), which accepts the initialization data through its arguments, and modifies the state of the object accordingly. Compared to the custom constructor, this has the advantage of not requiring a temporary copy of the object to be made. However, the calling syntax is (slightly) less convenient.

In the listing below, we demonstrate how to define the procedures to support both initialization mechanisms:

```
7   module Vec2D_class
8     implicit none
9     private ! Make module-entities "private" by default.
10
11    type, public :: Vec2D ! DT explicitly declared "public"
12      private ! Make internal data "private" by default.
13      real :: mU = 0., mV = 0.
14    contains
15      private ! Make methods "private" by default.
16      procedure, public :: init => initVec2D
17      ! . . . more methods (omitted in this example) . . .
18    end type Vec2D
19
```

---

[40] Making methods `private` can be useful, for example, when some of them are implementation-specific, and the users do not need to know about them.

[41] As we will demonstrate in the code, they cannot be type-bound procedure because they are functions which return the new DT-instance as their result.

```
20    ! Generic IFACE, for type-overloading
21    ! (to implement user-defined CTOR)
22    interface Vec2D
23       module procedure createVec2D
24    end interface Vec2D
25
26  contains
27    type(Vec2D) function createVec2D( u, v )  ! CTOR
28       real, intent(in) :: u, v
29       createVec2D%mU = u
30       createVec2D%mV = v
31    end function createVec2D
32
33    subroutine initVec2D( this, u, v )  ! init-subroutine
34       class(Vec2D), intent(inout) :: this
35       real, intent(in) :: u, v
36       ! copy-over data inside the object
37       this%mU = u
38       this%mV = v
39    end subroutine initVec2D
40  end module Vec2D_class
```

Listing 3.30 | src/Chapter3/dt_constructor_and_initializer.f90
(excerpt)

We can then use the module above, to declare and initialize values of type Vec2D, as follows:

```
42  program test_driver_b
43    use Vec2D_class
44    implicit none
45
46    type(Vec2D) :: A, D
47    ! ERROR: cannot define constants of DT with private data!
48    !type(Vec2D), parameter :: B = Vec2D(1.0, 3.2)
49    ! ERROR: cannot use user-CTOR to initialize at declaration!
50    !type(Vec2D) :: C = Vec2D(u=1.1, v=9.4)
51
52    ! Separate call to CTOR.
53    A = Vec2D(u=1.1, v=9.4)
54
55    ! Separate call to init-subroutine
56    call D%init(u=1.1, v=9.4)
57  end program
```

Listing 3.31 | src/Chapter3/dt_constructor_and_initializer.f90
(excerpt)

As demonstrated above,[42] user-defined constructors are limited (compared to the implicit constructor, which was provided by the compiler when direct data-access was allowed—Listing 3.27), as they can be used neither for defining constants based on a DT, nor for initializing an object on the same line where the object is declared—the only allowed use is to move the initialization outside the declarations part of the (sub)program.

As would be expected, these limitations also hold for the second initialization mechanism (using an "init" subroutine—see *line 56* above). Hence, the only benefits of custom constructors are the slightly more convenient syntax, and the fact that, being functions, they can be used directly in expressions. Otherwise, the choice of initialization mechanism depends on the preferences of the programmer (and, perhaps, project conventions).

However, we should emphasize that custom constructors can cause (depending on the compiler) unnecessary temporary objects to be created, which can degrade performance in some cases. Caution (and, even better, benchmarking) is advised

---

[42] Try to compile the code while un-commenting *line 48* and/or *50*.

when relying on custom constructors for large objects (e.g. those encapsulating model arrays in ESS), or when objects need to be repeatedly re-initialized, within time-consuming loops. We make use of both approaches in the code samples for the rest of the book.

A second problem with Listing 3.29, not solved by the updated DT-definition in Listing 3.30, was the lack of a mechanism for accessing the components of the vector. We can easily solve this, by adding two type-bound functions[43] (this time—*type-bound*), as shown below. These are also called *accessor*-methods (or *getters*, since their name is typically formed by concatenating "get" and the name of the component). Also, we re-introduce the getMagnitude-function:

```
 6  module Vec2d_class
 7    implicit none
 8    private  ! Make module-entities "private" by default.
 9
10    type, public :: Vec2d  ! DT explicitly declared "public"
11       private  ! Make internal data "private" by default.
12       real :: mU = 0., mV = 0.
13    contains
14       private  ! Make methods "private" by default.
15       procedure, public :: init => initVec2d
16       procedure, public :: getU => getUVec2d
17       procedure, public :: getV => getVVec2d
18       procedure, public :: getMagnitude => getMagnitudeVec2d
19    end type Vec2d
20
21    ! Generic IFACE, for type-overloading
22    ! (to implement user-defined CTOR)
23    interface Vec2d
24       module procedure createVec2d
25    end interface Vec2d
26
27  contains
28    type(Vec2d) function createVec2d( u, v )  ! CTOR
29       real, intent(in) :: u, v
30       createVec2d%mU = u
31       createVec2d%mV = v
32    end function createVec2d
33
34    subroutine initVec2d( this, u, v )  ! init-subroutine
35       class(Vec2d), intent(inout) :: this
36       real, intent(in) :: u, v
37       ! copy-over data inside the object
38       this%mU = u
39       this%mV = v
40    end subroutine initVec2d
41
42    real function getUVec2d( this )  ! accessor-method (GETter)
43       class(Vec2d), intent(in) :: this
44       getUVec2d = this%mU  ! direct-access IS allowed here
45    end function getUVec2d
46
47    real function getVVec2d( this )  ! accessor-method (GETter)
48       class(Vec2d), intent(in) :: this
49       getVVec2d = this%mV
50    end function getVVec2d
51
52    real function getMagnitudeVec2d( this ) result(mag)
53       class(Vec2d), intent(in) :: this
54       mag = sqrt( this%mU**2 + this%mV**2 )
55    end function getMagnitudeVec2d
56  end module Vec2d_class
```

Listing 3.32 | src/Chapter3/dt_accessors.f90 | (excerpt)

The DT can now be used similarly to the public-version (but note the change from A%mU to A%getU(), and same for v):

---

[43] Optimizing compilers should "see through" this intermediate layer, and inline the functions, so that they do not affect performance (although this needs to be verified through benchmarks, as usual).

```
67    ! Accessing components of DT through methods (type-bound procedures).
68    write(*, '(3(a,1x,f0.3))') "A%U =", A%getU(), &
69         ", A%V =", A%getV(), ", A%magnitude =", A%getMagnitude()
```

Listing 3.33 `src/Chapter3/dt_accessors.f90` (excerpt)

Exercise 12 (*Implementing SETters for a data type*) Complete the previous
example, by implementing procedures (name these `setU` and `setV`), to allow
users to individually modify the components of the type `Vec2d`.
Hint:
the procedures need to be implemented as subroutines, taking two arguments.

3.3.3 Inheritance (type Extension) and Aggregation

We mentioned in the beginning of Sect. 3.3 that the OOP paradigm usually leads to
a hierarchy of types. Two mechanisms are at the disposal of the Fortran programmer
to construct these hierarchies: *inheritance* and *aggregation*. We briefly discuss these
in this section.

As a simple showcase example, we will look at how to extend the DT from the
previous section (`Vec2d`), to represent 3*D*-vectors.[44]

3.3.3.1 Inheritance

A first mechanism for implementing hierarchies of types is *inheritance* (known as
"type-extension" in the Fortran standard). This achieves code-reuse by expressing
relations of the type "is a" between the entities modelled by the types. For example,
in a vegetation model in ESS, we may define a type `plant`, to collect attributes
relevant to our model for all plant types (e.g. albedo). Specialized types could then
be implemented for `tree`, `grass`, etc., which *inherit* basic characteristics from
`plant`, but also add some of their own (using Fortran terminology, we say that
`tree` *extends* `plant`). We also say that `plant` is the *parent/ancestor* type of
`tree` (or, equivalently, that `tree` is a *child/descendant* of `plant`). This process
of specialization could be continued, of course (by creating sub-classes for different
species of trees, etc.), although it is recommended not to make the hierarchy too
"tall" (i.e. to have too many levels of inheritance).

Returning to our simple example, here is how we could use inheritance to define
`Vec3d`:

---

[44] Of course, for such a simple DT, it would be easier (and potentially also more efficient) to write the
class for 3*D*-vectors from scratch. However, we implement it here based on `Vec2d`, to illustrate
the techniques in a simple setting.

```
49  module Vec3d_class
50      use Vec2d_class
51      implicit none
52      private
53
54      type, public, extends(Vec2d) :: Vec3d
55          private
56          real :: mW = 0.
57      contains
58          private
59          procedure, public :: getW => getWVec3d
60          procedure, public :: getMagnitude => getMagnitudeVec3d
61      end type Vec3d
62
63      interface Vec3d
64          module procedure createVec3d
65      end interface Vec3d
66
67  contains
68      ! Custom CTOR for the child-type.
69      type(Vec3d) function createVec3d( u, v, w )
70          real, intent(in) :: u, v, w
71          createVec3d%Vec2d = Vec2d( u, v ) ! Call CTOR of parent.
72          createVec3d%mW = w
73      end function createVec3d
74
75      ! Override method of parent-type.
76      ! (to compute magnitude, considering 'w' too)
77      real function getMagnitudeVec3d( this ) result(mag)
78          class(Vec3d), intent(in) :: this
79          ! this%Vec2d%getU() is equivalent, here, with this%getU()
80          mag = sqrt( this%Vec2d%getU()**2 + this%getV()**2 + this%mW**2 )
81      end function getMagnitudeVec3d
82
83      ! Method specific to the child-type.
84      ! (GETter for new component).
85      real function getWVec3d( this )
86          class(Vec3d), intent(in) :: this
87          getWVec3d = this%mW
88      end function getWVec3d
89  end module Vec3d_class
```

Listing 3.34 `src/Chapter3/dt_composition_inheritance.f90` (excerpt)

There are several remarkable points related to inheritance:

- The parent type needs to be stated with an `extends(<ParentTypeName>)` specifier (*line 54*)
- Inheritance automatically gives the child type a member of parent type, named after the parent. In our case, this is made clear by the call to the parent's custom constructor[45] (*line 71*). We get to this component, using the usual %-notation. Therefore, `createVec3d%Vec2d` (*line 71*) and `this%Vec2d` (*line 80*) are objects of type `Vec2d` in their own right. The child type can also directly access `public` data and methods of the parent (in our case, there was no `public` data, but we access the inherited methods `getU` and `getV` on *line 80*).
- It is possible to override (in the child type) methods of the parent.[46] We used this to define a new version of `getMagnitude` (*lines 75–81*), which correctly takes into account the additional component w. However, note that *while overriding, the interface of the method needs to remain constant* (except for the type of the passed-object dummy argument, which clearly needs to be different). For example,

---

[45] If the data of the parent was `public`, an *implicit constructor* would have been created for the child type, which would accept as arguments first the components of the parent type (in sequence), followed by the additional components of the child type (also in sequence).

[46] Unless those methods have the `non_overridable`-specifier in their binding attribute list.

we would not be allowed to override `getMagnitude` with a function that takes two arguments instead of one (assuming we would want that).

Our new DT could be used, as in:

```
91   program test_driver_inheritance
92      use Vec3d_class
93      implicit none
94      type(Vec3d) :: X
95
96      X = Vec3d( 1.0, 2.0, 3.0 )
97      write(*, '(4(a,f6.3))') "X%U =", X%getU(), ", X%V =", X%getV(), &
98           ", X%W =", X%getW(), ", X%magnitude =", X%getMagnitude()
99   end program test_driver_inheritance
```

Listing 3.35 | `src/Chapter3/dt_composition_inheritance.f90` | (excerpt)

In closing our quick coverage of inheritance note that, in Fortran jargon, the `class`-keyword indicates "class of types" (or inheritance hierarchy). This is different from other OOP languages, where "class" means a data type (`type` in Fortran). Also, unlike other languages, Fortran does not allow multiple inheritance (Metcalf et al. [8]).

3.3.3.2 Aggregation

A second mechanism for implementing hierarchies of types, which may come more natural to some programmers, is *aggregation*, which models a "has a" relationship between the types. We could also use this approach to implement another version of our `Vec3d`-class:

```
54      type, public :: Vec3d
55         private
56         type(Vec2d) :: mVec2d ! DT-aggregation
57         real :: mW = 0.
58      contains
59         private
60         procedure, public :: getU => getUVec3d
61         procedure, public :: getV => getVVec3d
62         procedure, public :: getW => getWVec3d
63         procedure, public :: getMagnitude => getMagnitudeVec3d
64      end type Vec3d
```

Listing 3.36 | `src/Chapter3/dt_composition_aggregation.f90` | (excerpt)

This is nothing else than simply using the less complex type as a component (*line 56*). The usual access-control mechanisms specify what data and methods of `Vec2d` can be referenced in the implementation of `Vec3d` (except that we now have to use the component's name, `mVec2d`, to get access). Since the implementation has no other remarkable features, we omit discussion of the methods here.

3.3.3.3 Inheritance or Aggregation?

The attentive reader may notice that the distinction of "is a" and "has a" relationships between DTs can sometimes be subjective. Indeed, to follow our previous example, the same type `Vec3d` was implemented with the same functionality based on either approach. This can make it confusing to select between the two in practice. A rough rule of thumb is to use inheritance if there is an obvious hierarchy of types in the problem, which will make children's direct inheritance of parent methods beneficial (no need to re-implement them, or to define "wrapper methods"). If, however, children would routinely need to override methods of the parent (or, worse, if parent methods do not make sense for children types!), aggregation is preferred as a composition method (see Rouson et al. [10]).

3.3.4 Procedure Overloading

Another important technique in OOP is *procedure overloading* (also known as "ad-hoc polymorphism"). Here, the idea is that several procedures can be accessed via the same name, and the compiler determines which one should be called, based on the types of their dummy arguments (also known as "signature"[47]). Clearly, for this to work, it is necessary that the two procedures actually have distinct signatures. To avoid confusion, note that this is different from *generic programming* (where a *unique* procedure definition is written by the programmer, and the compiler generates actual, callable procedures from this template where necessary—see Sect. 3.4). In *overloading*, the programmer is the one who will create the distinct functions for specific signatures explicitly.

To associate procedures to the same name for overloading, we need to define a *generic interface*, which we already encountered earlier while discussing access-control and information hiding with derived types (there, we needed to define a custom DT-constructor[48]). These are *named*[49] interface-blocks, where the name of the block will yield the name under which the overloads will be accessed. Inside generic interfaces, we can specify interfaces for *external procedures* by simply copying the relevant portions from the procedure definitions. However, for procedures defined in the same module as the generic interface, we need to use a module procedure <nameOfModuleProcedure>. Both cases are illustrated in the following example, which groups an external subroutine swapReal and a module procedure swapInteger, to make them callable via the generic name swap:

```
12  module Utilities
13    implicit none
14    private      ! Make things 'private' by default...
15    public swap  ! ...BUT, expose the generic-interface.
16    ! Generic interface
17    interface swap
18      ! Need explicit interface for non-module procedures...
19      subroutine swapReal( a, b )
20        real, intent(inout) :: a, b
21      end subroutine swapReal
22      ! ...BUT, module-procedures are attached with a
```

---

[47] This is essentially what we referred to as the "interface", without the return type, since most languages (including Fortran) do not look at this type when distinguishing overloads.

[48] This is also known as "type overloading", since the name of the generic interface was that of the type.

[49] Note that they serve a different purpose than the *unnamed* interface-blocks, shown in the beginning of this chapter (which were demonstrated for making the interface of an external procedure explicit).

```
23        ! 'module procedure'-statement.
24        module procedure swapInteger
25      end interface swap
26   contains
27      ! Module-procedure.
28      subroutine swapInteger( a, b )
29        integer, intent(inout) :: a, b
30        integer :: tmp
31        tmp = a; a = b; b = tmp
32      end subroutine swapInteger
33   end module Utilities
```

Listing 3.37 | `src/Chapter3/overload_normal_procedures.f90` | (excerpt)

The user of the module `Utilities` can then swap both `integers` and `reals`, using the same syntax:

```
35   program test_util_a
36     use Utilities
37     implicit none
38     integer :: i1 = 1, i2 = 3
39     real    :: r1 = 9.2, r2 = 5.6
40
41     write(*,'("Initial state:",1x,2(a,i0,1x), 2(a,f0.2,1x))') &
42           "i1 =", i1, ", i2 =", i2, ", r1 =", r1, ", r2 =", r2
43     call swap( i1, i2 )
44     call swap( r1, r2 )
45     write(*,'("State after swaps:",1x,2(a,i0,1x), 2(a,f0.2,1x))') &
46           "i1 =", i1, ", i2 =", i2, ", r1 =", r1, ", r2 =", r2
47   end program test_util_a
```

Listing 3.38 | `src/Chapter3/overload_normal_procedures.f90` | (excerpt)

Note that we can still access `swapReal` (even if it is `private`), through the generic interface (which is `public`).

In addition to the requirements that the overloads should have distinct signatures, note that they should also be all `functions` or all `subroutines`. Finally, it is also worth noting that there is an additional overloading mechanism for types, using what are known as "generic type-bound procedures". This is beneficial especially when the `only`-modifier is present at the place where `modules` are included (to import only selected entities). A mistake which can easily occur then is forgetting to include a generic interface, which can cause implicit functions (such as the assignment operator) to be called instead of the intended overloads in the module. We do not develop this issue (see Metcalf et al. [8] for details, if you encounter this scenario).

Operator overloading It is interesting to note that operators (like the unary $\boxed{\texttt{.not.}}$ or the binary $\boxed{+}$) are also procedures, only with special support from the language, to allow a more convenient notation (*infix notation*)—so the idea of overloading should apply to them as well. Indeed, Fortran (and other languages) allows developers to overload these functions for non-intrinsic types. We can simply achieve this by replacing the name of the generic interface ("swap" in our previous example) by $\boxed{\texttt{operator(<operatorName>)}}$, where `operatorName` is one of the intrinsic operators. This is demonstrated below:

```
8    module Vec3d_class
9      implicit none
10
11     type, public :: Vec3d
12        real :: mU = 0., mV = 0., mW = 0. ! Make 'private' in practice!
13      contains
14        procedure :: display ! Convenience output-method.
15     end type Vec3d
16
17     ! Generic interface, for operator-overloading.
```

```
18    interface operator(-)
19        module procedure negate      ! unary-minus
20        module procedure subtract    ! binary-subtraction
21    end interface operator(-)
22
23 contains
24    type(Vec3d) function negate( inVec )
25       class(Vec3d), intent(in) :: inVec
26       negate%mU = -inVec%mU
27       negate%mV = -inVec%mV
28       negate%mW = -inVec%mW
29    end function negate
30
31    ! NOTE: it is also possible to overload binary operators with heterogeneous
32    ! data-types. In our case, we could devine two more overloads for
33    ! binary-'-', to support subtraction when inVec1 or inVec2 is a scalar. In that
34    ! case, only the type of inVec1 or inVec2 needs to change, and the code inside
35    ! the function to be adapted.
36    type(Vec3d) function subtract( inVec1, inVec2 )
37       class(Vec3d), intent(in) :: inVec1, inVec2
38       subtract%mU = inVec1%mU - inVec2%mU
39       subtract%mV = inVec1%mV - inVec2%mV
40       subtract%mW = inVec1%mW - inVec2%mW
41    end function subtract
42
43    ! Utility-method, for more convenient display of 'Vec3d'-elements.
44    ! NOTE: A better solution is to use I/O for derived-types (see Metcalf2011).
45    subroutine display( this, nameString )
46       class(Vec3d), intent(in) :: this
47       character(len=*), intent(in) :: nameString
48       write(*,'(2a,3(f0.2,2x),a)') &
49             trim(nameString),"= (", this%mU, this%mV, this%mW, ")"
50    end subroutine display
51 end module Vec3d_class
```

Listing 3.39 | `src/Chapter3/overload_intrinsic_operators.f90` (excerpt)

The new operators can then be used to form expressions with our DT, as in:

```
53 program test_overload_intrinsic_operators
54    use Vec3d_class
55    implicit none
56    type(Vec3d) :: A = Vec3d(2., 4., 6.), B = Vec3d(1., 2., 3.)
57
58    write(*,'(/,a)') "initial-state:"
59    call A%display("A"); call B%display("B")
60
61    A = -A
62    write(*,'(/,a)') 'after operation "A = -A":'
63    call A%display("A"); call B%display("B")
64
65    A = A - B
66    write(*,'(/,a)') 'after operations "A = A - B":'
67    call A%display("A"); call B%display("B")
68 end program test_overload_intrinsic_operators
```

Listing 3.40 | `src/Chapter3/overload_intrinsic_operators.f90` (excerpt)

A constraint to be observed when overloading operators is that `functions` need to be used as actual procedures, which take one argument for unary operators, and two for binary operators respectively (where arguments have `intent(in)` in both cases).

Interestingly, it is even possible in Fortran to implement new (unary/binary) operators, which are not specified by the language standard. The syntax is similar to the previous case, except that we replace the name of the intrinsic operator with the desired name for our new operator (in the generic interface). For example, here is the interface block for a new operator `.cross.`, to compute the cross product of two vectors of type `Vec3d`:

```
18    ! Generic interface, for operator-overloading.
19    interface operator(.cross.)
20        module procedure cross_product  ! binary
21    end interface operator(.cross.)
```

Listing 3.41 | `src/Chapter3/overload_custom_operator.f90` (excerpt)

This powerful technique can lead to more readable code, by raising the level of abstraction, as in:

```
49        C = A .cross. B
```

Listing 3.42 `src/Chapter3/overload_custom_operator.f90` (excerpt)

Related to precedence, user-defined *unary* operators have higher priority than all other operators, while user-defined *binary* operators are the opposite (lowest priority—intrinsic operators included in both cases). However, it is easy (and often clearer) to override the order of evaluations with brackets, as usual.

Finally, another operator which can be overloaded is the assignment ($=$).[50] This is relevant only when the DT has a `pointer`-component, which is a topic outside the scope of this text.[51]

3.3.5 Polymorphism

Another OOP concept, related to inheritance, is *polymorphism*[52] ("many forms" in literal translation). The main characteristic of polymorphism is that entities may operate on data of different types, *but the type itself is dynamically resolved at runtime*. To support this concept, we can distinguish between:

- *polymorphic variables*: These are variables which may hold instances of different DTs during the execution of the program. They are used while implementing polymorphic procedures, and also for defining advanced data structures, such as a linked list (see Cormen et al. [6]) which may store different types of data in different nodes. Such variables may be defined in Fortran using the `class(<BaseClassName>)` or `class(*)` type.

The former allows the variable to be assigned a value of type `BaseClassName`, or any type which "is a" (=inherits from) `BaseClassName` (in Fortran jargon, we say that the variable is in the `class BaseClassName`). As in other OOP languages, it is possible to define the base class as `abstract`, so that variables of that type cannot be instantiated. Either way, the main purpose of the base type is to group common functionality, to be supported by all DT in the Fortran `class` (="inheritance hierarchy").

---

[50] This type of overloading is named "defined assignment", marked by changing the name of the generic interface by `assignment(=)` and implemented by a `subroutine` which takes two arguments (first `intent(out/inout)`) and second `intent(in)`).

[51] In that case, the implicit assignment implemented by the compiler would only perform a *shallow copy* of the object, without duplicating the data accessed by the `pointer`. However, when pointers are not used, the implicit assignment will perform a proper *deep copy* of the object, even when the DT has allocatable arrays as data members.

[52] To be precise, the concept we are referring to here is also known as "subtype polymorphism", to distinguish it from other methodologies which are also named "polymorphism" sometimes—e.g., *overloading* ("ad-hoc polymorphism") and *generic programming* ("parametric polymorphism").

When variables are defined with type `class(*)`, they can be assigned values of any DT (including intrinsic ones).

Due to their dynamic nature, polymorphic variables need to be `allocatable`, *dummy arguments*, or `pointers`.

- *polymorphic procedures*: These may operate on data of different types during the execution of the program. The advantage is that the code for such procedures can be written in generic terms, calling methods for variables of different DTs. As long as the DTs satisfy some interface conventions (the calls made by the polymorphic procedure need to actually exist in the callee's DT), the runtime system will dynamically determine the method of which DT needs to be called. In Fortran, polymorphic procedures are supported by using polymorphic variables (see above) as dummy arguments. It is also possible to take different actions based on the type of the actual arguments, using the `select type` -construct (which then supports matching a specific DT, or a `class` of DTs).

A more complete description of the mechanisms of polymorphism is outside the scope of this book. For more information, see Metcalf et al. [8] or Clerman and Spector [5].

3.4 Generic Programming (GP)

Languages like C++ also support GP, whereby procedures are written *once*, in terms of types that are specified later—see, e.g., Stepanov and McJones [11]. These can significantly reduce duplication of code; for example, a single `swap`-procedure can be written, from which the compiler may instantiate versions to swap data of `integer`, `real`, or user-defined type. Currently, Fortran also supports some of these ideas, but in a more limited sense.[53]

`elemental` **procedures** First, procedures can be made generic with respect to their rank, by making them `elemental`. Such functions take an array of any rank (including rank 0, so they also support scalars), and return an array of the same shape, but where each element in the output array contains the result of the function application to the corresponding element in the input array. When such an element-wise application makes sense, it can bring a significant reduction in code size (since it is not necessary to write specific versions of the procedure, for each array shape that may be used in our application). The following example demonstrates how this may be used with a `Vec3d` type, to implement vector normalization[54]:

---

[53] There is no "template metaprogramming" in Fortran.

[54] We left the components of the DT `public` here, for brevity.

```fortran
1   module Vec3d_class
2     implicit none
3     private
4     public :: normalize ! Expose the elemental function.
5
6     type, public :: Vec3d
7       real :: mU = 0., mV = 0., mW = 0.
8     end type Vec3d
9
10  contains
11    type(Vec3d) elemental function normalize( this )
12      type(Vec3d), intent(in) :: this
13      ! Local variable (note that the 'getMagnitude'-method could also be called,
14      ! but we do not have it implemented here, for brevity).
15      real :: magnitude
16      magnitude = sqrt( this%mU**2 + this%mV**2 + this%mW**2 )
17      normalize%mU = this%mU / magnitude
18      normalize%mV = this%mV / magnitude
19      normalize%mW = this%mW / magnitude
20    end function normalize
21  end module Vec3d_class
22
23  program test_elemental
24    use Vec3d_class
25    implicit none
26
27    type(Vec3d) :: scalarIn, array1In(10), array2In(15, 20)
28    type(Vec3d) :: scalarOut, array1Out(10), array2Out(15, 20)
29
30    ! Place some values in the 'in'-variables...
31    scalarOut = normalize( scalarIn ) ! Apply normalize to scalar
32    array1Out = normalize( array1In ) ! Apply normalize to rank-1 array
33    array2Out = normalize( array2In ) ! Apply normalize to rank-2 array
34  end program test_elemental
```

Listing 3.43 | `src/Chapter3/dt_elemental_normalization.f90`

Writing procedures as `elemental` not only make them generic, but can also improve performance. The latter is due to the fact that `elemental` procedures *are also required to be* `pure` (a topic we described in Sect. 3.2.5); with this restriction satisfied, it is guaranteed that the correct result will be obtained, no matter in which order (serial/parallel) the function is applied to the input elements. Many intrinsic procedures were designed to be `elemental`.

Parameterized types It is also possible[55] in Fortran to parameterize data types based on `integer`-values. Specific values for these parameters can then be assigned either at *compile-time* (also known as `kind`-like parameters, since they can be used to change the precision for the intrinsic types[56]), or at *runtime* (also known as `len`-like parameters, to highlight the connection with character strings of length assigned at runtime). For a discussion of this more advanced feature see e.g. Metcalf et al. [8].

References

1. Akin, E.: Object-Oriented Programming via Fortran 90/95. Cambridge University Press, Cambridge (2003)
2. Booch, G., Maksimchuk, R.A., Engle, M.W., Young, B.J., Connallen, J., Houston, K.A.: Object-Oriented Analysis and Design with Applications. Addison-Wesley Professional, Boston (2007)
3. Chapman, S.J.: Fortran 95/2003 for Scientists and Engineers. McGraw-Hill Science/Engineering/Math, New York (2007)

---

[55] Although this is standard Fortran 2003, most compilers had yet to implement this feature at the time of our writing unfortunately.

[56] Remember that `kind`-parameters are also `integer`-values.

4. Chivers, I.D., Sleightholme, J.: Compiler support for the Fortran 2003 and 2008 standards (Revision 11). ACM SIGPLAN Fortran Forum **31**(3), 17–28 (2012)
5. Clerman, N.S., Spector, W.: Modern Fortran: Style and Usage. Cambridge University Press, Cambridge (2011)
6. Cormen, T.H., Leiserson, C.E., Rivest, R.L., Stein, C.: Introduction to Algorithms. MIT Press, Cambridge (2009)
7. Hager, G., Wellein, G.: Introduction to High Performance Computing for Scientists and Engineers. CRC Press, Boca Raton (2010)
8. Metcalf, M., Reid, J., Cohen, M.: Modern Fortran Explained. Oxford University Press, Oxford (2011)
9. Richardson, L.F., Lynch, P.: Weather Prediction by Numerical Process. Cambridge University Press, Cambridge (2007)
10. Rouson, D., Xia, J., Xu, X.: Scientific Software Design: The Object-Oriented Way. Cambridge University Press, Cambridge (2011)
11. Stepanov, A.A., McJones, P.: Elements of Programming. Addison-Wesley Professional, Boston (2009)

Chapter 4
Applications

In the previous chapters, we kept the difficulty of the examples at a minimum, since the goal was to clearly illustrate basic Fortran features. Computer models in ESS are, of course, orders of magnitude more complex ($\sim 10^5$–10^6 lines of code are not uncommon). While it would not be practical (nor immediately instructive) to confront the readers directly with such a model, we attempt to make a "transition" towards more complex applications in this chapter, by presenting three case studies based on problems relevant to ESS. We start with a *finite differences* (FD) solver for the time-dependent heat diffusion problem in $2D$. The second case study (which is also more specific to ESS) discusses an EBM for simulating some feedbacks occurring in the climate system. Finally, we discuss a classic flow problem (*Rayleign-Bénard* (RB) convection), along with a numerical solution in $2D$ based on the *lattice Boltzmann method* (LBM).

4.1 Heat Diffusion

As a first application, we consider heat diffusion in $2D$. The governing equation for this phenomenon (in isotropic media) is:

$$\partial_t \theta = \kappa \partial_{\beta\beta} \theta \stackrel{\text{Not}}{\equiv} \kappa \nabla^2 \theta \tag{4.1}$$

where θ is temperature (in K) and κ is the thermal diffusivity coefficient (in m^2/s).

Index notation

For partial derivatives, we used the subscript notation, i.e. $\partial_{\beta\gamma} F(x, y) \stackrel{\text{Not}}{\equiv}$ $\frac{\partial^2 F(x,y)}{\partial x_\gamma \partial x_\beta}$ (for a function F depending on x and y). Also, x, y and t represent

---

© Springer-Verlag Berlin Heidelberg 2015
D.B. Chirila and G. Lohmann, *Introduction to Modern Fortran
for the Earth System Sciences*, DOI 10.1007/978-3-642-37009-0_4

the usual space and time coordinates. To keep the equations compact, we use the Einstein convention whenever possible, whereby a repeated subscript implies summation over the range of values of that subscript.

Mathematically, Eq. (4.1) is a second-order, parabolic *partial differential equation* (PDE), which we will solve numerically, with some initial and boundary conditions (acronyms ICs and BCs are used, respectively). Specifically, assume we are looking for the time-dependent temperature field within a solid square plate (assumed two-dimensional) with side length L, assuming that the temperature profile along the domain boundary is given by the following expressions:

$$\theta(x, L, t) = \frac{\theta_A - \theta_B}{L}x + \theta_B, \quad \theta(0, y, t) = \frac{\theta_B - \theta_C}{L}y + \theta_C, \quad (4.2)$$

$$\theta(x, 0, t) = \frac{\theta_D - \theta_C}{L}x + \theta_C, \quad \theta(L, y, t) = \frac{\theta_A - \theta_D}{L}y + \theta_D, \quad (4.3)$$

The previous expressions were derived by setting the temperature values at the four corners $(\theta_A, \ldots, \theta_D$—see also Fig. 4.1), and assuming the temperature along the edges to vary linearly between the values at the corresponding corners.

As IC for the interior (i.e. excluding the domain boundaries), we take:

$$\theta(x, y, 0) = \theta_A \quad (4.4)$$

with $\{x, y, t\} \in [0, L] \times [0, L] \times [0, \infty)$.

As concrete values, let us take $\theta_A = 100\,°C$, $\theta_B = 75\,°C$, $\theta_C = 50\,°C$, and $\theta_D = 25\,°C$, $L = 30\,m$ as the side length of the domain, and sandstone as the material $\kappa \approx 1.15 \times 10^{-6}\,m^2/s$.

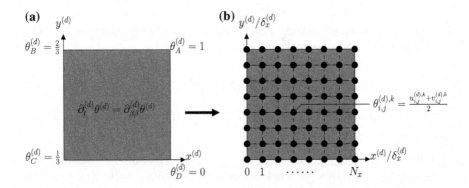

Fig. 4.1 Geometry for the $2D$ heat diffusion problem: **a** dimensionless system and **b** discretized (FD) system

4.1.1 Formulation in the Dimensionless System

Before presenting the procedure for solving our problem numerically, it is useful to re-write the equations in *dimensionless* form. This leverages the concept of *dynamical similarity*, as introduced by Buckingham [3]. The goal of this transformation is to minimize the number of parameters necessary to describe our physical problem. This makes our subsequent numerical solution more generally applicable, and also facilitates comparisons with analytical solutions, experimental data, or other numerical results.

> **Notation for distinguishing systems of units**
> Since several systems of units appear in the subsequent discussion, it is useful to introduce some conventions, to easily identify which of the systems any particular variable is associated with:
>
> - *physical system*: for consistency with the usual literature conventions, quantities measured in this system are written normally, without any superscript
> - *dimensionless system*: quantities measured in this system are typed with the "d" superscript ($x^{(d)}$, $t^{(d)}$, etc.)
> - *numerical system*: when reporting quantities in this system, they are typed with the "n" superscript ($x^{(n)}$, $t^{(n)}$, etc.)

Nondimensionalization consists of choosing some characteristic scales for the system under consideration, and rescaling the variables of the problem using these scales. For our current case, it is natural to use L as a characteristic length, and $\theta_A - \theta_D$ as a characteristic temperature difference. The formulation of the problem does not readily provide such a reference point for time, but we can define one as L^2/κ (also known as "diffusive" time scale). With these choices, we can define scaling relations for the primary variables:

$$x^{(d)} \equiv \frac{1}{L}x \iff x = Lx^{(d)} \tag{4.5}$$

$$t^{(d)} \equiv \frac{\kappa}{L^2}t \iff t = \frac{L^2}{\kappa}t^{(d)} \tag{4.6}$$

$$\theta^{(d)} \equiv \frac{\theta - \theta_D}{\theta_A - \theta_D} \iff \theta = \theta_D + (\theta_A - \theta_D)\theta^{(d)}, \tag{4.7}$$

and similar ones for the derivatives in our physical model:

$$\partial_t^{(d)} = \frac{L^2}{\kappa}\partial_t \iff \partial_t = \frac{\kappa}{L^2}\partial_t^{(d)} \tag{4.8}$$

$$\partial_{\alpha\beta}^{(d)} = L^2\partial_{\alpha\beta} \iff \partial_{\alpha\beta} = \frac{1}{L^2}\partial_{\alpha\beta}^{(d)} \tag{4.9}$$

Plugging equations (4.5)–(4.9) into Eq. (4.1), we obtain the heat diffusion equation in the nondimensional system of units:

$$\partial_t^{(d)} \theta^{(d)} = \partial_{\beta\beta}^{(d)} \theta^{(d)}, \tag{4.10}$$

which does not contain any transport coefficient anymore.

To complete the formulation of the problem, we also need to write the *boundary conditions* (BCs) and IC in nondimensional units. Using Eq. (4.7), Eqs. (4.2), (4.3) become:

$$\theta^{(d)}(x^{(d)}, 1, t^{(d)}) = \frac{\theta_A - \theta_B}{\theta_A - \theta_D} x^{(d)} + \frac{\theta_B - \theta_D}{\theta_A - \theta_D} = \frac{1}{3} x^{(d)} + \frac{2}{3} \tag{4.11}$$

$$\theta^{(d)}(0, y^{(d)}, t^{(d)}) = \frac{\theta_B - \theta_C}{\theta_A - \theta_D} y^{(d)} + \frac{\theta_C - \theta_D}{\theta_A - \theta_D} = \frac{1}{3} y^{(d)} + \frac{1}{3} \tag{4.12}$$

$$\theta^{(d)}(x^{(d)}, 0, t^{(d)}) = \frac{\theta_D - \theta_C}{\theta_A - \theta_D} x^{(d)} + \frac{\theta_C - \theta_D}{\theta_A - \theta_D} = -\frac{1}{3} x^{(d)} + \frac{1}{3} \tag{4.13}$$

$$\theta^{(d)}(1, y^{(d)}, t^{(d)}) = y^{(d)} \tag{4.14}$$

and corresponding to Eq. (4.4) we have:

$$\theta^{(d)}(x^{(d)}, y^{(d)}, 0) = 1 \tag{4.15}$$

4.1.2 Numerical Discretization of the Problem

To obtain a numerical solution of Eq. (4.10) (*continuous*), we need a system of *discretized* equations, which can be solved algebraically with a computer. Several general approaches can be used for this—for example, *finite differences* (FD), *finite volumes* (FV), or *finite elements* (FE). In the present case, we use FD, whereby the temperature field is represented only at discrete points within the spatial domain (see Fig. 4.1b), and also only at discrete time steps.

The link to the continuous equations is established by approximating the partial derivatives with algebraic expressions, depending on temperature values on the grid. Many formulae can be constructed for approximating the partial derivatives in the models, which differ in *accuracy* and *computational cost* (there is usually a trade-off between these two factors). Another important aspect for time-dependent problems is *stability*, since certain combinations of FD discretizations can lead to unstable (or conditionally stable) systems of algebraic equations.

We refer the interested reader to Pletcher et al. [20] or Strang [27] for more details on discretization and numerical analysis. For our purposes here, we use a specific FD-discretization method for Eq. (4.1), which was first proposed by Barakat and Clark [1]. It belongs to the class of *alternating-direction explicit* (ADE) methods, and is unconditionally stable, 2nd-order accurate[1] in space and time, and also easy to implement (but with some disadvantages in terms of parallel execution, as we will discuss in next chapter). The idea is to average the solutions obtained with two specially designed discretizations of Eq. (4.10), so that the leading-order errors of each discretization are of similar magnitude and opposite sign to those of the other discretization—leading to a higher accuracy of the combined solution.

Notation for discretized equations
It is common to use subscripts for the discretized variables, to mark the fact that they refer to values at a certain location in the discretized domain. Similarly, we use superscripts to denote a specific time step. With this notation, $\theta_{i,j}^{(d),\,k}$ is the dimensionless temperature at location $(x^{(d)} = i\delta_x^{(d)};\ y^{(d)} = j\delta_y^{(d)})$ and time $t^{(d)} = k\delta_t^{(d)}$.

With the notation explained above, the two discretizations for the method of Barakat and Clark [1] read:

$$\frac{u_{i,j}^{(d),\,k+1} - u_{i,j}^{(d),\,k}}{\delta_t^{(d)}} = \frac{1}{\left(\delta_x^{(d)}\right)^2}\left[(u_{i-1,j}^{(d),\,k+1} - u_{i,j}^{(d),\,k+1} - u_{i,j}^{(d),\,k} + u_{i+1,j}^{(d),\,k}) \right.$$
$$\left. + (u_{i,j-1}^{(d),\,k+1} - u_{i,j}^{(d),\,k+1} - u_{i,j}^{(d),\,k} + u_{i,j+1}^{(d),\,k}) \right]$$
$$(4.16)$$

$$\frac{v_{i,j}^{(d),\,k+1} - v_{i,j}^{(d),\,k}}{\delta_t^{(d)}} = \frac{1}{\left(\delta_x^{(d)}\right)^2}\left[(v_{i-1,j}^{(d),\,k} - v_{i,j}^{(d),\,k} - v_{i,j}^{(d),\,k+1} + v_{i+1,j}^{(d),\,k+1}) \right.$$
$$\left. + (v_{i,j-1}^{(d),\,k} - v_{i,j}^{(d),\,k} - v_{i,j}^{(d),\,k+1} + v_{i,j+1}^{(d),\,k+1}) \right]$$
$$(4.17)$$

where $\delta_t^{(d)} \ll 1$ is the dimensionless time between any two successive iterations and $\delta_x^{(d)} = \delta_y^{(d)} \ll 1$ (*isotropic* grid) is the dimensionless distance between adjacent nodes. The final discrete solution for temperature is obtained by averaging the sub-solutions:

---

[1] Here, accuracy refers to the difference between the solution of the continuous equation and that of the discretized equation. This "discretization error" (also known as "truncation error") occurs due to the method itself. The magnitude of this error is expected to decrease for a well-constructed discretized scheme as the number of grid points is increased (the order of the scheme quantifies how fast this error decreases). An additional type of errors ("roundoff error") is introduced by the fact that digital computers can only store most real numbers with a limited accuracy, as we discussed in Sect. 2.3.4.

$$\theta_{i,j}^{(d),\,k} = \frac{u_{i,j}^{(d),\,k} + v_{i,j}^{(d),\,k}}{2} \tag{4.18}$$

With some additional notations, i.e.:

$$\lambda \equiv \frac{2\delta_t^{(d)}}{(\delta_x^{(d)})^2} \tag{4.19}$$

$$A \equiv \frac{1 - \lambda}{1 + \lambda} \tag{4.20}$$

$$B \equiv \frac{\lambda}{2(1 + \lambda)}, \tag{4.21}$$

we can cast the algebraic equations into a more convenient form for implementation:

$$u_{i,j}^{(d),\,k+1} = A u_{i,j}^{(d),\,k} + B\left[\left(u_{i-1,j}^{(d),\,k+1} + u_{i+1,j}^{(d),\,k}\right) + \left(u_{i,j-1}^{(d),\,k+1} + u_{i,j+1}^{(d),\,k}\right)\right] \tag{4.22}$$

$$v_{i,j}^{(d),\,k+1} = A v_{i,j}^{(d),\,k} + B\left[\left(v_{i-1,j}^{(d),\,k} + v_{i+1,j}^{(d),\,k+1}\right) + \left(v_{i,j-1}^{(d),\,k} + v_{i,j+1}^{(d),\,k+1}\right)\right] \tag{4.23}$$

The equations above are, in principle, implicit (since we have some remaining terms on the *right-hand side* (RHS), which refer to variables at time step $k+1$). The "trick" is to notice that, if we adopt a particular order of node updates for $u^{(d)}$ (progressively advancing grid nodes from time k to $k+1$), the missing terms will be available "for free", since they correspond to locations which were already brought to the new time step. The same observation holds for $v^{(d)}$ too, only that the reverse node ordering has to be used. These aspects will be important for our later Fortran implementation.

The computational costs (in terms of memory and computing time), as well as the accuracy of the numerical solution, are dictated by the choice of the discretization parameters ($\delta_x^{(d)}$ and $\delta_t^{(d)}$). In programming practice, it is convenient to refer to the integer-valued parameters $N_x = 1/\delta_x^{(d)}$ and $N_t = 1/\delta_t^{(d)}$, representing the number of discrete space intervals necessary for representing a characteristic length, and similarly for time.[2] Unlike most explicit methods, implicit methods such as the one we use here have the advantage of remaining stable, for any combination of positive N_x and N_t. However, the (transient) numerical solution may not be physically meaningful if the discrete time step $\delta_t^{(d)}$ is too large. A safe choice (see Barakat and Clark [1]) is:

$$\delta_t^{(d)} = (\delta_x^{(d)})^2 \iff N_t = N_x^2 \tag{4.24}$$

The geometry for the problem is sketched in Fig. 4.1.

---

[2] Note that the number of discrete *points* used for representing the characteristic length is actually $N_x + 1$; similarly, there are $N_t + 1$ time steps (including the initial state at $t = 0$) for simulating the evolution of the system during a characteristic time duration.

4.1.3 Implementation (Using OOP)

We use the OOP-approach to implement this example, so we start by identifying the entities that would be involved in our program, and the functionality (methods) they should support. As mentioned in Sect. 3.3, this process is to a large degree subjective, so there is more than one acceptable solution for this task (see e.g. Robinson [24] and the references therein for more examples of software design).

module *NumericKinds*: As many other numerical algorithms, the present solver is sensitive to the range and accuracy of the numeric types used to represent the different model variables. Therefore, we use the procedure discussed in Sect. 3.2.7 for providing explicit requirements for these types. For all examples in this chapter, we group such declarations in the `NumericKinds`-module. This not only guarantees that equivalent kinds are selected even if the code is run on multiple vendor platforms, but also allows convenient switching of the precision of the variables globally. For the current application, this module reads:

```
module NumericKinds
  implicit none

  ! KINDs for different types of REALs
  integer, parameter :: &
    R_SP = selected_real_kind(  6,   37 ), &
    R_DP = selected_real_kind( 15,  307 ), &
    R_QP = selected_real_kind( 33, 4931 )
  ! Alias for precision that we use in the program (change this to any of the
  ! values 'R_SP', 'R_DP', or 'R_QP', to switch to another precision globally).
  integer, parameter :: RK = R_DP ! if changing this, also change RK_FMT

  ! KINDs for different types of INTEGERs
  integer, parameter :: &
    I1B = selected_int_kind(2),  & ! max = 127
    I2B = selected_int_kind(4),  & ! max ~ 3.28x10^4
    I3B = selected_int_kind(9),  & ! max ~ 2.15x10^9
    I4B = selected_int_kind(18)    ! max ~ 9.22x10^18
  ! Alias for integer-precision (analogue role to RK above).
  integer, parameter :: IK = I3B

  ! Edit-descriptors for real-values
  character(len=*), parameter :: R_SP_FMT = "f0.6", &
    R_DP_FMT = "f0.15", R_QP_FMT = "f0.33"
  ! Alias for output-precision to use in the program (keep this in sync with RK)
  character(len=*), parameter :: RK_FMT = R_DP_FMT
end module NumericKinds
```

Listing 4.1 | `src/Chapter4/solve_heat_diffusion_v1.f90` | (excerpt)

With this piece of "infrastructure" out of the way, we can develop a strategy for structuring our program implementation. Here, we propose a simple decomposition, consisting of a `Config` and a `Solver` type, each in its own `module`. Motivating this decomposition is the principle of decoupling parts of the program which are expected to undergo future modifications. In our case, it is sensible to decouple the code which reads the simulation parameters from the actual solver, because we will extend each of these components in Chap. 5, for demonstrating additional techniques (the `Config` type will be improved with `namelist`-support, while the `Solver` type will be extended with rudimentary support for parallel processing). In a large application, such a separation of the physical problem formulation from the numerical method would also open the possibility of switching solvers if this becomes necessary at a later stage in the project (in this case, it would probably also prove useful to further partition the `Solver` type itself).

`Config` **type:** From the FD-method discussed above, we can identify several parameters that are relevant to our solution. Because we will demonstrate two methods for reading these parameters into the program, it is useful to group this data in one place, as a distinct type. To avoid implementing too many methods of SET/GET-variety, we will leave the variables in the DT `public`.[3] There is no need for type-bound procedures in this DT, but it would be useful to provide a custom constructor, to initialize a `Config`-instance from a file on disk (see file `Chapter4/config_file_formatted.in` in the source code repository). The declarations part of the corresponding module and the procedure interfaces are shown below:

```
module Config_class
  use NumericKinds
  implicit none
  private

  type, public :: Config
     real(RK) :: mDiffusivity = 1.15E-6_RK, & ! sandstone
                 ! NOTE: "physical" units here (Celsius)
           mTempA = 100._RK, &
           mTempB =  75._RK, &
           mTempC =  50._RK, &
           mTempD =  25._RK, &
           mSideLength = 30._RK
     integer(IK) :: mNx = 200 ! # of points for square side-length
  end type Config

  ! Generic IFACE for user-defined CTOR
  interface Config
     module procedure createConfig
  end interface Config

contains
  type(Config) function createConfig( cfgFilePath )
     character(len=*), intent(in) :: cfgFilePath
     integer :: cfgFileID
     open( newunit=cfgFileID, file=trim(cfgFilePath), status='old', action='read' )
     close(cfgFileID)
  end function createConfig
end module Config_class
```

Listing 4.2 `src/Chapter4/solve_heat_diffusion_v1.f90` (excerpt)

`Solver` **type**: This type forms the core of our solution. As data members, it encapsulates an object of type `Config`, from which some other variables are evaluated:

- `mNt` (representing N_t) is assigned a value (i.e. $N_t = N_x^2$) in the `Solver` type, since such constraints are usually specific to the numerical algorithm used (a different method would most probably have different limitations)
- `mDx` and `mDt` are the discretization parameters, inversely proportional to N_x and N_t
- `mA` and `mB` are pre-factors (defined in Eqs. (4.19)–(4.21)) for expressing the algorithm more concisely
- `mNumItersMax` represents the total number of algorithm iterations to be performed

Also as data members, we have two dynamic arrays (`mU` and `mV`), which will hold the state of the two FD sub-solutions. To simplify things, we will not implement

---

[3] This does breach the OOP idea of encapsulation, but not in a significant way here, since this DT is essentially part of the implementation of the `Solver` DT (which does not expose it further).

a custom constructor for this type.[4] Finally, $\boxed{\texttt{mCurrIter}}$ keeps track of the current iteration number, to serve as documentation for the simulation output.

The methods for `Solver`, can be grouped into `public` and `private` ones:

- `public`: As far as the user code is concerned, it would be reasonable to add methods for: (1) initializing a `Solver` based on a user-specified file containing parameters (delegating the actual reading of the file to the data member of type `Config`), (2) performing the time marching and (3) writing an output file (whose name is to be specified by the user). To facilitate debugging, we also add a method to inquire the temperature at a certain position.

  Note that in this interface part of the *abstract data type* (ADT) we did not mention too many details specific to the actual numerical method used in the `Solver`—the only time when the user of this module interacts with the details of the method is while creating the configuration file (specifically, through the choice of N_x). This practice of keeping the interface as generic as possible is very natural with OOP, and can lead to more maintainable programs. Additionally, it allows a potential user of our data types to easily switch to a new `Solver`, if one becomes available.[5]

- `private`: To implement the method for time marching (run-subroutine), we can define two `private`-methods in the `Solver` DT, each one responsible for updating one of our two sub-solution fields. Also, we add a `final` (destructor) method, to demonstrate how these can be bound to the type.[6]

The declarations part of the corresponding module, procedure interfaces, and some parts of the implementations are shown below:

```
module Solver_class
  use NumericKinds
  use Config_class
  implicit none
  private

  type, public :: Solver
    private ! Hide internal-data from users.
    type(Config) :: mConfig
    real(RK) :: mNt, & ! # of iterations to simulate a characteristic time
      mDx, mDt, mA, mB ! Configuration-dependent factors.
    real(RK), allocatable, dimension(:,:) :: mU, mV ! main work-arrays
    integer(IK) :: mNumItersMax, mCurrIter = 0
  contains
    private ! By default, hide methods (and expose as needed).
    procedure, public :: init
    procedure, public :: run
    procedure, public :: writeAscii
    procedure, public :: getTemp
    ! Internal methods (users don't need to know about these).
    procedure :: advanceU
    procedure :: advanceV
    !final :: cleanup ! NOTE: may need to comment-out for gfortran!
  end type Solver

contains
  subroutine init( this, cfgFilePath, simTime ) ! initialization subroutine
    class(Solver), intent(inout) :: this
    character(len=*), intent(in) :: cfgFilePath
```

---

[4] A custom constructor could lead to unnecessary copying of the `allocatable` arrays (which can often be large) when initializing the object with assignments between `Solver`-instances. This problem could be circumvented in principle by making $\boxed{\texttt{mU}}$ and $\boxed{\texttt{mV}}$ pointers, but it would also obscure the present discussion.

[5] Equivalently, it allows the implementer of the `Solver` type to improve the internal implementation, without affecting the user code dramatically.

[6] Note that this feature was still not implemented in several compilers at the time of this writing (notably—$\boxed{\texttt{gfortran-4.8}}$).

```
      real(RK), intent(in) :: simTime
      this%mV = this%mU
end subroutine init

real(RK) function getTemp( this, i, j ) ! GETter for temperature
   class(Solver), intent(in) :: this
   integer(IK), intent(in) :: i, j
   getTemp = 0.5*( this%mU(i,j) + this%mV(i,j) )
end function getTemp

subroutine run( this ) ! method for time-marching
   class(Solver), intent(inout) :: this
   integer(IK) :: k ! dummy index (time-marching)

   do k=1, this%mNumItersMax ! MAIN loop
      ! simple progress-monitor
      if( mod(k-1, (this%mNumItersMax-1)/10) == 0 ) then
         write(*, '(i5,a)') nint((k*100.0)/this%mNumItersMax), "%"
      end if
      ! defer work to private methods
      call this%advanceU()
      call this%advanceV()
      this%mCurrIter = this%mCurrIter + 1 ! tracking time step
   end do
end subroutine run

subroutine advanceU( this )
   class(Solver), intent(inout) :: this
   integer(IK) :: i, j ! local variables
   ! actual update for 'mU'-field (NE-ward)
   do j=1, this%mConfig%mNx-1    ! do NOT update
      do i=1, this%mConfig%mNx-1 ! boundaries
         this%mU(i,j) = this%mA*this%mU(i,j) + this%mB*( &
               this%mU(i-1,j) + this%mU(i+1,j) + this%mU(i,j-1) + this%mU(i,j+1) )
      end do
   end do
end subroutine advanceU

subroutine advanceV( this ) ! similar to 'advanceU'
   class(Solver), intent(inout) :: this
   ! ...............................................
   end do
end subroutine advanceV

! method for producing a ASCII output file
subroutine writeAscii( this, outFilePath )
   class(Solver), intent(in) :: this
   character(len=*), intent(in) :: outFilePath
   ! ...............................................
   integer(IK) :: x, y, outFileID ! local variables

   open( newunit=outFileID, file=trim(outFilePath), status='replace', action='write' )
   close(outFileID)
end subroutine writeAscii

! destructor method
subroutine cleanup( this )
   ! 'class' -> 'type' (dummy-arg cannot be polymorphic for final procedures)
   type(Solver), intent(inout) :: this
   ! in this version, we only deallocate memory
   deallocate( this%mU, this%mV )
end subroutine cleanup
end module Solver_class
```

Listing 4.3 `src/Chapter4/solve_heat_diffusion_v1.f90` (excerpt)

Note that subroutines `advanceU` and `advanceV` scan the domain in opposite directories, updating their corresponding arrays "in-place" (i.e. while the update is in progress, these arrays will contain values at both time step n and $n+1$).

With the types defined above, we can write a very compact main-program:

```
program solve_heat_diffusion_v1
   use NumericKinds
   use Solver_class
   implicit none

   type(Solver) :: square
   real(RK) :: simTime = 0.1 ! no. of characteristic time-intervals to simulate

   character(len=200) :: configFile = "config_file_formatted.in", &
         outputFile = "simulation_final_temp_field.dat"

   call square%init( configFile, simTime ) ! call Initializer
   call square%run()

   call square%writeAscii( outputFile )
end program solve_heat_diffusion_v1
```

Listing 4.4 `src/Chapter4/solve_heat_diffusion_v1.f90` (excerpt)

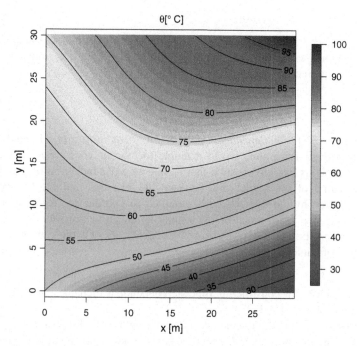

θ[° C]

Fig. 4.2 Plot of the numerical solution for the transient heat diffusion equation at $t^{(d)} = 0.1$, using the method of Barakat and Clark [1]

A sample solution, obtained with the parameter values listed in the declaration of the `Config` type and with the main-program shown above and time step dictated by Eq. (4.24) is given in Fig. 4.2 (the plot was produced with the script `src/Chapter4/plotHeatDiffSoln.R`, also in the source code repository).

> **Exercise 13** (*Heat diffusion in a rectangular domain*) Extend the code provided in the repository (file `solve_heat_diffusion_v1.f90`), so that it works for rectangular domains too (not only for squares). Use your program to simulate a different choice of temperatures at the boundaries.

> **Exercise 14** (*Robust code with error checking*) In the interest of clarity, the example presented in this section did not include error checking for the cases when the input and output files cannot be opened, or when the configuration data cannot be read, due to mistakes in the input file. Extend our example (or your modified version produced for Exercise 13), to increase the robustness

of the code (by checking the `status`-argument in `open`/`close`-statements, and by adding exception statement labels for the `read`/`write`-statements, as discussed in Sect. 2.4.3).

Exercise 15 (*Time-dependent output for heat diffusion*) Extend the `writeAscii`-subroutine in our example (or one of your versions created in the previous exercises), such that a time step-dependent suffix is added to the name of the output file, between the filename and the file extension selected by the user. Use this modified subroutine, to create output at each time step, and visualize the results as an animation (modifying the script `src/Chapter4/plotHeatDiffSoln.R`, or using your visualization tool of choice).

Hints:

- A procedure for constructing time step-dependent filenames was shown at the end of Sect. 2.4.3.
- Be sure to estimate the amount of data that would be produced on disk— adjust the number of simulation time steps, to keep the output reasonable.

We will revisit this example in Sect. 5.2.1 (to improve the methodology for specifying input parameters), and in Sect. 5.3.5 (to show how to improve performance slightly with parallelization).

4.2 Climate Box Model

Here we provide a brief overview of an inter-hemispheric box model of the deep ocean circulation to study the feedbacks in the climate system. The model is based on an ocean box model of the Atlantic Ocean [25] coupled to an energy balance model of the atmosphere [17, 21]. The inter-hemispheric box model consists of four oceanic and three atmospheric boxes, as indicated in Fig. 4.3. The ocean boxes represent the Atlantic Ocean from 80°N to 60°S, with a width of 80° (assumed constant). The indices of the temperatures T, the salinities S, the surface heat fluxes H, the atmospheric heat fluxes F, the radiation terms R as well as later on the volumes bear on the different boxes ($\Box_N$ for the northern, $\Box_M$ for the tropical, $\Box_D$ for the deep and $\Box_S$ for the southern box).

For simplicity, the discrete boxes are assumed to be homogeneous, i.e. temperatures and the salinities everywhere within one box are alike. The climate model is based on mass and energy considerations. Emphasis is placed on the overturning flow Φ of the ocean circulation.

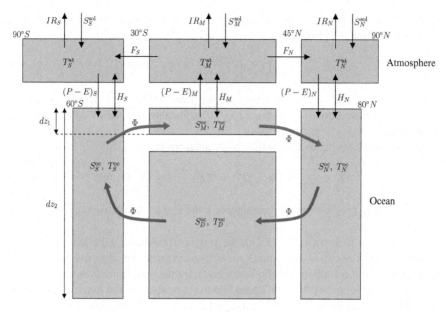

Fig. 4.3 Schematic illustration of the climate box model

The prognostic equations for the temperatures of the ocean boxes consist of two parts. The first part is proportional to the overturning flow Φ and represents the advective (transport) coupling between the boxes. The second part, which is dependent on the surface heat flux H, stands for the coupling between the ocean and the atmosphere. This latter part is missing for the deep box, which is not connected to the atmosphere. The four differential equations for the ocean temperatures read:

$$\frac{d}{dt}T_S^{oc} = -\left(T_S^{oc} - T_D^{oc}\right)\frac{\Phi}{V_S} + \frac{H_S}{\rho_0 c_p dz_2}, \tag{4.25}$$

$$\frac{d}{dt}T_M^{oc} = -\left(T_M^{oc} - T_S^{oc}\right)\frac{\Phi}{V_M} + \frac{H_M}{\rho_0 c_p dz_1}, \tag{4.26}$$

$$\frac{d}{dt}T_N^{oc} = -\left(T_N^{oc} - T_M^{oc}\right)\frac{\Phi}{V_N} + \frac{H_N}{\rho_0 c_p dz_2} \quad \text{and} \tag{4.27}$$

$$\frac{d}{dt}T_D^{oc} = -\left(T_D^{oc} - T_N^{oc}\right)\frac{\Phi}{V_D} \tag{4.28}$$

where ρ_0 denotes a reference density for saltwater and c_p the specific heat capacity of water. The depths of the discrete ocean boxes were chosen as $dz_M = dz_1 = 600\,\text{m}$ and $dz_N = dz_S = dz_2 = 4{,}000\,\text{m}$; volumes of the boxes are denoted by $V_{i,\ i\in\{N,M,S,D\}}$.

The overturning flow is assumed to be proportional to the density gradients of the oceanic boxes after Stommel [26]. Like in Rahmstorf [23] the northern and the southern box will be taken into account for this, which leads to the equation for the calculation of the overturning flow:

$$\Phi = c \left[-\alpha \left(T_N^{\text{oc}} - T_S^{\text{oc}} \right) + \beta \left(S_N^{\text{oc}} - S_S^{\text{oc}} \right) \right] \tag{4.29}$$

where the constants α and β represent the thermal and the haline expansion coefficients in the equation of state; c is an adjustable parameter which is set to produce present day overturning rates.

The surface heat fluxes follow the equations from Haney [15]:

$$H_i = Q_{1,i} - Q_{2,i} \left(T_i^{\text{oc}} - T_i^{\text{at}} \right), \quad i \in \{S, M, N\} \tag{4.30}$$

where $Q_{1,i}$ and $Q_{2,i}$ are tuning parameters for the surface heat fluxes (a pair for each atmosphere box).

Analogously to Eqs. (4.27), (4.28), the prognostic differential equations for the salinities consist of two components. One of those is again the advective part, caused by the interconnection between the boxes and the other one quantifies the effects of the freshwater fluxes between the ocean and the atmosphere. The latter is again only for the boxes near the surface, thus the equations are:

$$\frac{d}{dt} S_S^{\text{oc}} = - \left(S_S^{\text{oc}} - S_D^{\text{oc}} \right) \frac{\Phi}{V_S} - S_{\text{ref}} \frac{(P-E)_S}{dz_2}, \tag{4.31}$$

$$\frac{d}{dt} S_M^{\text{oc}} = - \left(S_M^{\text{oc}} - S_S^{\text{oc}} \right) \frac{\Phi}{V_M} + S_{\text{ref}} \frac{(P-E)_M}{dz_1}, \tag{4.32}$$

$$\frac{d}{dt} S_N^{\text{oc}} = - \left(S_N^{\text{oc}} - S_M^{\text{oc}} \right) \frac{\Phi}{V_N} - S_{\text{ref}} \frac{(P-E)_N}{dz_2}, \tag{4.33}$$

$$\frac{d}{dt} S_D^{\text{oc}} = - \left(S_D^{\text{oc}} - S_N^{\text{oc}} \right) \frac{\Phi}{V_D}. \tag{4.34}$$

The reference salinity S_{ref} is a characteristic average value for the entire Atlantic Ocean, and the freshwater fluxes are denoted as precipitation minus evaporation $(P - E)$. These freshwater fluxes are calculated by the divergence of the latent heat transport in the atmosphere and are assumed to be proportional to the meridional moisture gradient explained below.

The atmospheric EBM calculates the heat fluxes between the ocean and atmosphere, as well as horizontal latent and sensible heat transports as diffusion following Chen et al. [5]. The EBM contains sensible and latent heat transports, radiation R_i, as well as the surface heat fluxes H_i between the atmosphere and the ocean. The atmospheric temperatures T_i^{at} follow the prognostic equations:

$$c_2 \frac{d}{dt} T_S^{at} = \frac{\partial \left(F_S^s + F_S^l \right)}{\partial y} + R_S - H_S,$$ (4.35)

$$c_2 \frac{d}{dt} T_M^{at} = - \frac{\partial \left(F_S^s + F_N^s + F_S^l + F_N^l \right)}{\partial y} + R_M - H_M,$$ (4.36)

$$c_2 \frac{d}{dt} T_N^{at} = \frac{\partial \left(F_N^s + F_N^l \right)}{\partial y} + R_N - H_N.$$ (4.37)

where c_2 is related to the specific heat of air. The sensible F_i^s and latent F_i^l heat transport are described in dependence of the meridional gradient of the surface temperature and moisture q_i^s, respectively:

$$F_i^s = K_s \frac{\partial T_i^{at}}{\partial y}$$ (4.38)

$$F_i^l = K_l \frac{\partial q_i^s}{\partial y} .$$ (4.39)

with $i \in \{S, N\}$. K_s and K_l are empirical parameters, which must be adjusted to generate realistic values for the sensible and latent heat transports. The radiation terms R_i in Eqs. (4.35)–(4.37) consist of an incoming solar shortwave S_i and an outgoing infrared longwave I_i part. The extraterrestrial solar radiation is not absorbed entirely, and a latitude-dependent average albedo α_i is introduced to account for the reflectance. The outgoing infrared radiation I_i is calculated through a linear formula of Budyko [4]. Thus, the equation for the net radiation balance is

$$R_i = S_i - I_i = S_i^{sol} (1 - \alpha_i) - \left(A + B T_i^{at} \right).$$ (4.40)

The model calculates the freshwater fluxes from the divergence of the latent heat transport, assuming a proportionality of the form:

$$(P - E)_i \sim \partial F_i^l / \partial y$$ (4.41)

4.2.1 Numerical Discretization

The theoretical model presented above ultimately leads to a system of *coupled coupled ordinary differential equations* (ODEs). Since the spatial dependence was incorporated into the choice and properties of the discrete model boxes, we only need to consider dependence on time. Specifically, we need to choose a discretization for the time derivatives in the *left-hand side* (LHS) of the governing Eqs. (4.25)–(4.28), (4.31)–(4.37). To keep the example simple, we use the Euler forward scheme[7] for

---

[7] For Exercise 17, we briefly discuss how to extend the code with a more accurate integration scheme.

this, whereby the general equation:

$$\frac{d}{dt}X_i(t) = f_i(\mathbf{X}(t)) \tag{4.42}$$

is discretized as:

$$\frac{X_i^{k+1} - X_i^k}{\delta_t} \approx f_i(\mathbf{X}^k), \tag{4.43}$$

which can be written as:

$$X_i^{k+1} \approx X_i^k + \delta_t \cdot f_i(\mathbf{X}^k) \tag{4.44}$$

where we have, in our case:

$$\mathbf{X} = (T_S^{\text{oc}}, T_M^{\text{oc}}, T_N^{\text{oc}}, T_D^{\text{oc}}, S_S^{\text{oc}}, S_M^{\text{oc}}, S_N^{\text{oc}}, S_D^{\text{oc}}, T_S^{\text{at}}, T_M^{\text{at}}, T_N^{\text{at}})^T \tag{4.45}$$

and X_i denotes any of these variables, and f_i—the accompanying expression on the RHS of the evolution equations.

In the model, we use a time step $\delta_t = 1/100\,yr$, to ensure the stability of the system according to the *Courant-Friedrichs-Levy* (CFL) criterion [8].[8]

4.2.2 Implementation (OOP/SP Hybrid)

We use a combination of OOP and SP techniques to implement this example. As a first step, we identify some basic entities in the model and, for those which we will group as classes, the methods they need to support. These will provide the foundation on which we will eventually build the climate box model. The complete example is in the file box_model_euler.f90 , in the source code repository.

module *NumericKinds*: Similar to the previous example, we use a module to specify requirements for the numeric kinds (see Sect. 4.1.3 for details).
module *PhysicsConstants*: Our model will use quite a few physical and time-related constants. It is easier to group these in a module of their own, which looks like:

```
module PhysicsConstants
  use NumericKinds
  implicit none
  public

  real(RK), parameter :: &
       RHO_SEA_WATER = 1025.,      & ! [kg/m^3]
  ! .................................
       WIDTH_ATLANTIC = 80. ! lateral span of the Atlantic [degrees of longitude]
end module PhysicsConstants
```

Listing 4.5 src/Chapter4/box_model_euler.f90 (excerpt)

---

[8] For an English translation, refer to [9].

Since we do not expect these to change, we do not need to create a separate type to encapsulate these constants, so a plain `module` should suffice.

module *ModelConstants*: In addition to the physics constants, several model-dependent parameters will also appear, to control aspects of numerics (e.g. time step), or for tuning parameterizations used by the model. Normally, these should be encapsulated into a separate type, to allow reading them from a file. However, to simplify things, we do not show this here, and create another plain `module` instead[9]:

```fortran
module ModelConstants
  use NumericKinds
  use PhysicsConstants
  implicit none
  public

  real(RK), parameter :: &
      NO_YEARS = 10000., & ! total simulation time [yr]
      DT_IN_YEARS = 1./100., & ! time-step [yr]
      DTS = DT_IN_YEARS * SECONDS_IN_YEAR, & ! time-step [s]
      ! tuning-parameters for surface heat-fluxes
      Q1_S = 10., Q2_S = 50., &
      ! .................................
      ! volumes of the ocean boxes
      V_S = AREA_S*DZ2, V_M = AREA_M*DZ1, V_N = AREA_N*DZ2, V_D = AREA_D*(DZ2-DZ1)

  integer, parameter :: &
      NO_T_STEP = int(NO_YEARS / DT_IN_YEARS), & ! number of model-iterations
      OUTPUT_FREQUENCY = 100
end module ModelConstants
```

Listing 4.6 `src/Chapter4/box_model_euler.f90` (excerpt)

module *GeomUtils*: When the heat and freshwater fluxes are expressed in spherical coordinates (not discussed above for brevity—see e.g. Nakamura et al. [19] and Prange et al. [21] for details), several trigonometric expressions appear, which depend on the latitude bounds of the model boxes, given in degrees. On the other hand, the corresponding Fortran intrinsic functions operate with radians. Therefore, to keep the model formulation compact, we provide two functions (`rad2Deg` and `deg2Rad`) for converting between these units, and we also write variants of the necessary trigonometric functions, which take angles measured in degrees as input arguments (`sinD` and `cosD`). Another geometric function (`linInterp`), which we bundle in the same module, provides basic linear interpolation, for estimating the latent heat fluxes between the atmospheric model boxes. The resulting module is shown below:

```fortran
module GeomUtils
  use NumericKinds
  use PhysicsConstants
  implicit none
  public

contains
  ! Convert radians to degrees
  real(RK) function rad2Deg( radians )
    real(RK), intent(in) :: radians
    rad2Deg = radians / ONE_DEG_IN_RADS
  end function rad2Deg

  ! Convert degrees to radians
  real(RK) function deg2Rad( degrees )
    real(RK), intent(in) :: degrees
    deg2Rad = degrees * ONE_DEG_IN_RADS
  end function deg2Rad

  ! Sine of an angle given in degrees
  real(RK) function sinD( degrees )
    real(RK), intent(in) :: degrees
    sinD = sin( deg2Rad( degrees ) )
  end function sinD
```

---

[9] See Sect. 4.1.3 for an example of encapsulating configuration data, and Sect. 5.2.1 for improving that further by using a `namelist`.

```
! Cosine of an angle given in degrees
real(RK) function cosD( degrees )
  real(RK), intent(in) :: degrees
  cosD = cos( deg2Rad( degrees ) )
end function cosD

! Estimate y(x), given 2 points (x0,y0) & (x1,y1).
real(RK) function linInterp( x0, y0, x1, y1, x )
  real(RK), intent(in) :: x0, y0, x1, y1, x
  ! Check for malformed-denominator in division.
  ! NOTE: More robust option is to use IEEE features of
  ! Fortran 2003 (see e.g. Clerman 2011).
  if( abs(x1-x0) < (epsilon(x)*abs((x - x0))) ) then
    stop("FP error in linInterp-function! Aborting.")
  end if
  ! Actual computation
  linInterp = y0 + (x - x0)/(x1 - x0)*(y1 - y0)
end function linInterp
end module GeomUtils
```

Listing 4.7 `src/Chapter4/box_model_euler.f90` (excerpt)

`ModelState` **type**: From the description in the previous section, note that the model is conceptually just a system of coupled ODEs. We define the *abstract data type* (ADT) `ModelState` as a container for the state vector. We only add a few "methods" for this new type:

- `getCurrModelState` returns the model state for writing output.
- `preventOceanFreezing` prevents the model from entering physical regimes beyond its scope. In particular, our model does not account for potential sea ice formation. Therefore, if ocean temperatures decrease below the freezing point temperature, we issue a warning and bring them back to this value.[10]
- `computePhi` calculates the intensity of the overturning circulation. As for the previous procedure, we issue a warning if the overturning circulation seems to be reversed, since the model is not designed for that situation (same comment applies here also).

For a more expressive formulation of the numerical scheme (later, in the main program), we make extensive use of operator overloading (procedures `scalarTimesModelState` , `modelStateTimesScalar` , `addModelStates` and `subtractModelStates` [11]).

As "free" subroutines for the new ADT, we have `newModelState` , which constructs a new instance of the type (based on ICs of the model), `dQSdT` , which computes the slope of the saturation vapor pressure and, finally, `dModelState` . This last procedure is particularly important, since it encodes the actual physics of our model (the RHS of the evolution equations). The procedure also returns a ModelState-instance, representing the rate of change of the model state. Here, the fact that the procedure is *not* type-bound (and requires a ModelState input argument), since this facilitates evaluation of the rate of change at fractional

---

[10] Note that, for production code, it would probably be a better idea to stop the program altogether if such an exceptional situation occurs, to rule-out any misinterpretations (for various reasons, the warnings may not reach the user).

[11] The numerical schemes we use do not actually need this last subroutine, but we include it since it makes it easier to get the difference between two solutions.

time steps, as usually required by higher-order numerical integration schemes (see Exercise 17).

```fortran
module ModelState_class
  use iso_fortran_env, only : error_unit
  use PhysicsConstants
  use ModelConstants
  use GeomUtils
  implicit none

  type :: ModelState
    real(RK) :: TocS, TocM, TocN, TocD, SocS, SocM, SocN, SocD, TatS, TatM, TatN
  contains
    procedure, public :: getCurrModelState
    procedure, public :: preventOceanFreezing
    procedure, public :: computePhi
  end type ModelState

  interface ModelState
    module procedure newModelState
  end interface ModelState

  interface operator(*)
    module procedure scalarTimesModelState
    module procedure modelStateTimesScalar
  end interface operator(*)

  interface operator(+)
    module procedure addModelStates
  end interface operator(+)

  interface operator(-)
    module procedure subtractModelStates
  end interface operator(-)

contains
  ! Calculate the slope of saturation vapor pressure w.r.t temperature
  ! (see Rogers and Yau, Cloud Physics, 1976, p.16).
  ! NOTE: Unlike the other procedures in the module, this one is not type-bound.
  real(RK) function dQSdT( Tc )
    real(RK), intent(in) :: Tc
    real(RK) :: p, ex, sat
    p = 1000.
    ex = 17.67 * Tc/( Tc + 243.5 )
    sat = 6.112 * exp( ex )
    dQSdT = 243.5 * 17.67 * sat / (Tc + 243.5)**2 * 0.622 / p
  end function dQSdT

  ! User-defined CTOR (INITer).
  type(ModelState) function newModelState( TocS, TocM, TocN, TocD, &
       SocS, SocM, SocN, SocD, TatS, TatM, TatN ) result(new)
    real(RK), intent(in) :: TocS, TocM, TocN, TocD, &
         SocS, SocM, SocN, SocD, TatS, TatM, TatN
    ! ................................
  end function newModelState

  type(ModelState) function scalarTimesModelState( scalar, state ) result(res)
    real(RK), intent(in) :: scalar
    class(ModelState), intent(in) :: state
    ! ................................
  end function scalarTimesModelState

  type(ModelState) function modelStateTimesScalar( state, scalar ) result(res)
    class(ModelState), intent(in) :: state
    real(RK), intent(in) :: scalar
    ! re-use 'scalarTimesModelState'
    res = scalar*state
  end function modelStateTimesScalar

  type(ModelState) function addModelStates( state1, state2 ) result(res)
    class(ModelState), intent(in) :: state1, state2
    ! ................................
  end function addModelStates

  type(ModelState) function subtractModelStates( state1, state2 ) result(res)
    class(ModelState), intent(in) :: state1, state2
    ! ................................
  end function subtractModelStates

  ! "Brute-force" resetting of ocean-temperatures, if they decrease below
  ! freezing-point.
  subroutine preventOceanFreezing( this )
    class(ModelState), intent(inout) :: this

    if( this%TocS < TFREEZE_SEA_WATER) then
      this%TocS = TFREEZE_SEA_WATER
      write(error_unit, '(a)') "Warning: TocS was reset to prevent freezing!"
    end if
  end subroutine preventOceanFreezing

  real(RK) function computePhi( this ) result(phi)
    class(ModelState), intent(in) :: this

    phi = C*(-ALPHA*(this%TocN-this%TocS) + BETA*(this%SocN-this%SocS))
    if( phi < 0. ) then
      phi=0. ! prevent reversal of circulation
      write(error_unit, '(a)') "Warning: reversal of circulation detected!"
    end if
```

```fortran
      end function computePhi

      function getCurrModelState( this ) result(res)
        class(ModelState), intent(in) :: this
        real(RK) :: res(13)
        ! local vars
        real(RK) :: tempGlobal

        tempGlobal = (0.5*this%TatS+1.207*this%TatM+0.293*this%TatN)/2.

        res = [ tempGlobal, this%TocS, this%TocM, this%TocN, this%TocD, &
              this%SocS, this%SocM, this%SocN, this%SocD, &
              this%TatS, this%TatM, this%TatN, &
              this%computePhi()*1.E-6 ] ! units are transformed to [Sv]
      end function getCurrModelState

      ! Physics is encoded here (i.e. RHS of evolution equations)
      type(ModelState) function dModelState( old )
        type(ModelState), intent(in) :: old
        real(RK) :: F30S, F45N, phi, Tat30S, Tat45N, FsS, FsN, FlS, FlN, hS, hM, hN, &
              fwFaS, fwFaN, rS, rM, rN, midLatS, midLatM, midLatN

        midLatS = rad2Deg( asin( (sinD(LAT1_AT_S)+sinD(LAT2_AT_S))/2._RK ) )
        midLatM = rad2Deg( asin( (sinD(LAT1_AT_M)+sinD(LAT2_AT_M))/2._RK ) )
        midLatN = rad2Deg( asin( (sinD(LAT1_AT_N)+sinD(LAT2_AT_N))/2._RK ) )

        Tat30S = linInterp(x0=midLatM, y0=old%TatM, x1=midLatS, y1=old%TatS, x=-30._RK)
        Tat45N = linInterp(x0=midLatM, y0=old%TatM, x1=midLatN, y1=old%TatN, x= 45._RK)

        FsS = KS * (old%TatM-old%TatS)/(R_E*(deg2Rad(midLatM)-deg2Rad(midLatS)))
        FsN = KS * (old%TatM-old%TatN)/(R_E*(deg2Rad(midLatN)-deg2Rad(midLatM)))

        FlS = KL*RH*dQSdT(Tat30S) / (old%TatM - old%TatS) * FsS/KS
        FlN = KL*RH*dQSdT(Tat45N) / (old%TatM - old%TatN) * FsN/KS

        F30S = FsS + FlS
        F45N = FsN + FlN

        fwFaS = ( LR*2*PI*R_E*cosD(30._RK)*FlS ) * (80./360.)
        fwFaN = ( LR*2*PI*R_E*cosD(45._RK)*FlN ) * (80./360.) * 2.5

        hS = Q1_S - Q2_S*(old%TocS-old%TatS)
        hM = Q1_M - Q2_M*(old%TocM-old%TatM)
        hN = Q1_N - Q2_N*(old%TocN-old%TatN)
        ! radiation-balance terms
        rS = S_SOL_S*(1.-ALBEDO_S) - (A+B*old%TatS)
        rM = S_SOL_M*(1.-ALBEDO_M) - (A+B*old%TatM)
        rN = S_SOL_N*(1.-ALBEDO_N) - (A+B*old%TatN)

        phi = old%computePhi()

        ! Final phase: preparing the function-result
        !! Ocean Temperatures
        dModelState%TocS = -(old%TocS-old%TocD)*phi/V_S + hS/RCZ2
        dModelState%TocM = -(old%TocM-old%TocS)*phi/V_M + hM/RCZ1
        dModelState%TocN = -(old%TocN-old%TocM)*phi/V_N + hN/RCZ2
        dModelState%TocD = -(old%TocD-old%TocN)*phi/V_D
        !! Ocean Salinities
        dModelState%SocS = -(old%SocS-old%SocD)*phi/V_S - S_REF*fwFaS/V_S
        dModelState%SocM = -(old%SocM-old%SocS)*phi/V_M + S_REF*(fwFaS+fwFaN)/V_M
        dModelState%SocN = -(old%SocN-old%SocM)*phi/V_N - S_REF*fwFaN/V_N
        dModelState%SocD = -(old%SocD-old%SocN)*phi/V_D
        !! Atmosphere Temperatures
        dModelState%TatS = ( &
            (cosD(30._RK)*F30S) / (R_E*(sinD(90._RK)-sinD(30._RK))) &
            + rS - FRF_S*hS &
            )/( CP_DRY_AIR*BETA_S )

        dModelState%TatM = ( &
            -(cosD(30._RK)*F30S+cosD(45._RK)*F45N) / (R_E*(sinD(30._RK)+sinD(45._RK))) &
            + rM - FRF_M*hM &
            )/( CP_DRY_AIR*BETA_M )

        dModelState%TatN = ( &
            (cosD(45._RK)*F45N) / (R_E*(sinD(90._RK)-sinD(45._RK))) &
            + rN - FRF_N*hN &
            )/( CP_DRY_AIR*BETA_N )
      end function dModelState
```

Listing 4.8 | `src/Chapter4/box_model_euler.f90` | (excerpt)

main-program: With the pieces of "infrastructure" presented above, we can write our main-program. Here, we use an SP-approach, to highlight the algorithm better:

```fortran
program box_model_euler
  use PhysicsConstants
  use ModelConstants
  use ModelState_class
  implicit none

  integer :: i, outFileID
  type(ModelState) :: stateSim1E, statePerturbation

  ! Perturbation to superimpose over equilibrium state.
  statePerturbation = ModelState( TocS = 0., TocM = 0., &
      TocN = 0., TocD = 0., &
```

```
        SocS = 0., SocM = 0., &
        SocN = -0.7, SocD = 0., &
        TatS = 0., TatM = 0., TatN = 0. )

  stateSim1E = ModelState( TocS = 4.77740431, TocM = 24.42876625, &
        TocN = 2.66810894, TocD = 2.67598915, &
        SocS = 34.40753555, SocM = 35.62585068, &
        SocN = 34.92513657, SocD = 34.91130066, &
        TatS = 4.67439556, TatM = 23.30437851, TatN = 0.94061828) + statePerturbation

  ! prepare for output
  open(newunit=outFileID, file="box_model_euler.out", &
        form="formatted", status="replace")

  ! write initial conditions
  write(outFileID, '(14('//RK_FMT//', 1x))') 0._RK, stateSim1E%getCurrModelState()

  do i=1, NO_T_STEP
        ! Euler-forward step
        stateSim1E = stateSim1E + DTS*dModelState(stateSim1E)

        call stateSim1E%preventOceanFreezing()

        ! Conditional OUTPUT-writing
        if( mod(i-1, OUTPUT_FREQUENCY) == 0 ) then
           write(outFileID, '(14('// RK_FMT // ', 1x))') i*DT_IN_YEARS, &
                 stateSim1E%getCurrModelState()
        end if
  end do
  close(outFileID) ! Clean-up for output
end program box_model_euler
```

Listing 4.9 Main-program for box model application (see file `src/Chapter4/box_modelc _euler.f90`)

Exercise 16 (*Investigations with the box model*)

1. In the regions of deep water formation in the North Atlantic, relatively small amounts of fresh water added to the surface can stabilize the water column to the extent that convection can be prevented from occurring. Such interruption decreases the poleward mass transport Φ in the ocean. Furthermore, perturbations of the meridional transport in the ocean can be amplified by positive feedbacks: a weaker northward salt transport brings less dense water to high latitudes, which further reduces the meridional transport. Discuss the case where the initial conditions in salinity at latitudes is changed.

2. The coupled model shall be used to investigate the sensitivity of the system with respect to radiative forcing and stochastic weather perturbations. Additional radiative forcing may come from increased tracer gas concentrations in the atmosphere, whereas the atmospheric weather fluctuations may reflect unresolved effects of the atmospheric transports modeled as white noise.

3. The initial values of the model are chosen to represent present-day climate conditions. Determine which parameters in the model affect the overturning streamfunction most.

Exercise 17 (*4th-order Runge-Kutta integration for the box model*) In the code above, we used the Euler forward scheme for integrating the system of equations numerically, which has the advantage of simplicity. However, many alternative schemes exists, which have much higher accuracy than Euler forward. A popular scheme, for example, is the 4th-order Runkge-Kutta (also known as *RK4*—see Press et al. [22]), which replaces Eq. (4.44) by the following evolution equations:

$$X_i^{k+1} \approx X_i^k + \frac{\delta_t}{6} [k_1 + 2(k_2 + k_3) + k_4] \tag{4.46}$$

where:

$$k_1 = f_i \left(\mathbf{X}^k \right) \tag{4.47}$$

$$k_2 = f_i \left(\mathbf{X}^k + \frac{\delta_t}{2} k_1 \right) \tag{4.48}$$

$$k_3 = f_i \left(\mathbf{X}^k + \frac{\delta_t}{2} k_2 \right) \tag{4.49}$$

$$k_4 = f_i \left(\mathbf{X}^k + \delta_t k_3 \right) \tag{4.50}$$

Extend the program presented above with this discretization scheme, and compare the results with those obtained using the previous discretization.

We will revisit this example in Sect. 5.1.2.3 to illustrate how to spread the components of the project across distinct files, to make it more modular.

4.3 Rayleigh-Bénard (RB) Convection in 2*D*

As a last (and somewhat more involved) example, we consider the evolution of an incompressible fluid, with temperature coupled as a passive tracer, using the Boussinesq approximation. We will present a program which solves this problem numerically in 2*D*, using the *lattice Boltzmann method* (LBM). The geometry of the problem is sketched in Fig. 4.4. The fluid domain is bounded by two parallel horizontal planes. We assume that all gradients along the *Z*-direction (perpendicular to the page) are negligible, so that we can treat the problem as two-dimensional. While certainly restrictive, this assumption still permits the study of interesting physics, such as the transition from a stationary state to convective flow.

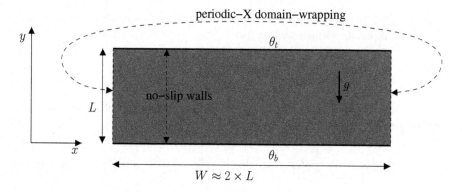

Fig. 4.4 Geometry for the 2D RB problem (see text for description)

4.3.1 Governing Equations

The evolution equations for the incompressible fluid and for the temperature component read:

$$\partial_\beta u_\beta = 0 \tag{4.51}$$

$$\rho\left(\partial_t u_\gamma + u_\beta \partial_\beta u_\gamma\right) = -\partial_\gamma p + \nu\rho\partial_{\epsilon\epsilon} u_\gamma + \rho r_\gamma, \qquad \forall \gamma \in \{x, y\} \tag{4.52}$$

$$\partial_t \theta + u_\beta \partial_\beta \theta = \kappa\partial_{\epsilon\epsilon}\theta \tag{4.53}$$

where the symbols represent:

- ρ—fluid density ($= \rho_0$ = const. for strictly incompressible fluids)
- u—fluid velocity
- T—fluid temperature
- r—acceleration due to any body forces; in our case: $r = -g\hat{y} \Leftrightarrow r_\gamma = -g\delta_{\gamma,y}$, where δ is the Kronecker delta
- ν—kinematic viscosity of the (Newtonian) fluid
- κ—thermal diffusivity
- subscripts β, γ and $\epsilon \in \{x, y\}$; repeated subscripts in the same term imply summation over these possible values of the indices.

In the Boussinesq approximation, the requirement of having a constant density is slightly relaxed: the fluid density is allowed to change in response to variations in temperature. However, these density variations are assumed to be important only when they appear multiplied by the gravity force. The dependence of density on temperature is linearized for simplicity (thus, the current model does not support salinity variations):

$$\rho = \rho_0\left[1 - \alpha(\theta - \theta_0)\right], \tag{4.54}$$

where α is the coefficient of thermal expansion, and ρ_0 is a reference density, at a reference temperature which we take as:

$$\theta_0 = \frac{\theta_b + \theta_t}{2}. \qquad (4.55)$$

With these assumptions, and also absorbing the constant part of the gravity force into the pressure term, Eq. (4.52) becomes:

$$\partial_t u_\gamma + u_\beta \partial_\beta u_\gamma = -\frac{1}{\rho_0}\partial_\gamma p + \nu \partial_{\epsilon\epsilon} u_\gamma + \alpha g(\theta - \theta_0)\delta_{\gamma,y}, \quad \forall \gamma \in \{x, y\} \quad (4.56)$$

where p now stands for the modified pressure.

BCs for the simulation (see also Fig. 4.4)

- at the horizontal walls

  – The velocity field is assumed to satisfy the no-slip condition:

  $$u_\gamma(x, 0, t) = u_\gamma(x, L, t) = 0, \quad \forall \gamma \in \{x, y\} \qquad (4.57)$$

  – Also, driving the flow is a vertical temperature gradient, imposed through the temperature BCs ($\theta_b > \theta_t$):

  $$\theta(x, 0, t) = \theta_b \equiv \theta_0 + \frac{\Delta\theta_0}{2} \qquad (4.58)$$

  $$\theta(x, L, t) = \theta_t \equiv \theta_0 - \frac{\Delta\theta_0}{2} \qquad (4.59)$$

  where we introduced:

  $$\Delta\theta_0 \equiv \theta_b - \theta_t \qquad (4.60)$$

- at the lateral walls we use periodic BCs, for simplicity:

  $$\theta(0, y, t) = \theta(W, y, t) \qquad (4.61)$$

  $$u_\gamma(0, y, t) = u_\gamma(W, y, t), \quad \forall \gamma \in \{x, y\} \qquad (4.62)$$

ICs

- The velocity field is set identically to zero everywhere initially:

  $$u_\gamma(x, y, 0) = 0, \quad \forall \gamma \in \{x, y\} \qquad (4.63)$$

- The initial temperature is given by a linear profile, which matches the values at the horizontal boundaries:

$$\theta(x, y, 0) = \theta_0 + \Delta\theta_0 \left(\frac{1}{2} - \frac{y}{L} \right) \tag{4.64}$$

This configuration actually represents a stable solution of the governing equations, under certain conditions (which we quantify later).

4.3.2 Problem Formulation in Dimensionless Form

Just as for the problem discussed in Sect. 4.1, we prefer to re-write the governing equations in a dimensionless form, instead of solving them directly in the physical system of units (in terms of meters, seconds, Kelvin, etc.). For the following calculations, we adopt the same notation conventions as explained in that section. For the RB problem, it is natural to choose the height of the channel (L) as the characteristic distance, which suggests the following scaling relations:

$$x^{(d)} \equiv \frac{x}{L} \iff x = Lx^{(d)} \tag{4.65}$$

Because the flow is initially at rest, we construct a characteristic time scale based on diffusivity[12] and the characteristic length. The resulting scaling relations for time read:

$$t^{(d)} \equiv \frac{\kappa}{L^2} t \iff t = \frac{L^2}{\kappa} t^{(d)} \tag{4.66}$$

Similarly, $\frac{\kappa}{L}$ can be chosen as a characteristic velocity, which leads to:

$$u_\gamma^{(d)} = \frac{L}{\kappa} u_\gamma \iff u_\gamma = \frac{\kappa}{L} u_\gamma^{(d)}, \quad \forall \gamma \in \{x, y\} \tag{4.67}$$

The characteristic (modified) pressure difference can be defined based on the characteristic velocity, i.e. $\rho_0 \frac{\kappa^2}{L^2}$, so that:

$$p^{(d)} = \frac{L^2}{\rho_0 \kappa^2} p \iff p = \frac{\rho_0 \kappa^2}{L^2} p^{(d)} \tag{4.68}$$

Finally, it is natural to scale the temperature differences based on $\Delta\theta_0$ and θ_0, defined earlier:

$$\theta^{(d)} = \frac{\theta - \theta_0}{\Delta\theta_0} \iff \theta = \theta_0 + \Delta\theta_0 \theta^{(d)} \tag{4.69}$$

---

[12] $[\kappa]_{SI} = m^2/s$; note that some authors use α instead of κ for denoting the thermal diffusivity.

Using the chain rule, the scaling relations for the derivatives can also be obtained:

$$\partial_\beta^{(d)} = L\partial_\beta \iff \partial_\beta = \frac{1}{L}\partial_\beta^{(d)} \tag{4.70}$$

$$\partial_{\beta\gamma}^{(d)} = L^2\partial_{\beta\gamma} \iff \partial_{\beta\gamma} = \frac{1}{L^2}\partial_{\beta\gamma}^{(d)} \tag{4.71}$$

$$\partial_t^{(d)} = \frac{L^2}{\kappa}\partial_t \iff \partial_t = \frac{\kappa}{L^2}\partial_t^{(d)} \tag{4.72}$$

Using these characteristic scales, the equations of conservation for fluid mass Eq. (4.51), momentum Eq. (4.56) and heat Eq. (4.53) become in the dimensionless system:

$$\partial_\beta^{(d)} u_\beta^{(d)} = 0 \tag{4.73}$$

$$\partial_t^{(d)} u_\gamma^{(d)} + u_\beta^{(d)} \partial_\beta^{(d)} u_\gamma^{(d)} = -\partial_\gamma^{(d)} p^{(d)} + \Pr \partial_{\epsilon\epsilon}^{(d)} u_\gamma^{(d)} + \mathrm{Ra}\ \Pr \theta^{(d)} \delta_{\gamma,y}\ , \quad \forall \gamma \in \{x,y\} \tag{4.74}$$

$$\partial_t^{(d)} \theta^{(d)} + u_\beta^{(d)} \partial_\beta^{(d)} \theta^{(d)} = \partial_{\epsilon\epsilon}^{(d)} \theta^{(d)} \tag{4.75}$$

where the dimensionless coefficients Ra (Rayleigh number) and Pr (Prandtl number) are defined as:

$$\Pr \equiv \frac{\nu}{\kappa} \tag{4.76}$$

$$\mathrm{Ra} \equiv \frac{\alpha g L^3 \Delta\theta_0}{\kappa\nu} \tag{4.77}$$

BCs in dimensionless form

- at the horizontal walls

  – velocity:

$$u_\gamma^{(d)}(x^{(d)}, 0, t^{(d)}) = u_\gamma^{(d)}(x^{(d)}, 1, t^{(d)}) = 0, \quad \forall \gamma \in \{x,y\} \tag{4.78}$$

  – temperature:

$$\theta^{(d)}(x^{(d)}, 0, t^{(d)}) = +\frac{1}{2} \tag{4.79}$$

$$\theta^{(d)}(x^{(d)}, 1, t^{(d)}) = -\frac{1}{2} \tag{4.80}$$

- at the lateral walls (periodic wrapping of the domain)

$$\theta^{(d)}(0, y^{(d)}, t^{(d)}) = \theta^{(d)}\left(\frac{W}{L}, y^{(d)}, t^{(d)}\right) \tag{4.81}$$

$$u_\gamma^{(d)}(0, y^{(d)}, t^{(d)}) = u_\gamma^{(d)}\left(\frac{W}{L}, y^{(d)}, t^{(d)}\right), \quad \forall \gamma \in \{x, y\} \tag{4.82}$$

ICs in dimensionless form

- velocity:

$$u_\gamma^{(d)}(x^{(d)}, y^{(d)}, 0) = 0, \quad \forall \gamma \in \{x, y\} \tag{4.83}$$

- temperature:

$$\theta^{(d)}(x^{(d)}, y^{(d)}, 0) = \frac{1}{2} - y^{(d)} \tag{4.84}$$

- pressure (chosen so that the pressure and buoyancy terms in Eq. (4.74) are in balance initially):

$$p^{(d)}(x^{(d)}, y^{(d)}, 0) = \frac{\text{Ra} \, \text{Pr}}{2} y^{(d)}(1 - y^{(d)}) \tag{4.85}$$

It is interesting to note that the dynamical behaviour of the system is determined by the coefficients Pr and Ra (see Eqs. (4.76) and (4.77)). For example, two geometrically similar setups **A** and **B** with $\Delta\theta_{0,A} L_A^3 = \Delta\theta_{0,B} L_B^3$ have identical solutions in the dimensionless system (the flows are said to be *dynamically similar*).[13]

As already mentioned, a possible state of the system corresponds to the fluid being at rest, with the temperature field undergoing pure diffusion (which results in a linear temperature profile, depending on the y-coordinate only). Linear stability theory (see Tritton [30] and references therein) predicts that this solution is realized as long as the Rayleigh number is

$$\text{Ra} < \text{Ra}_{\text{crit.},1} = 1,707.762 \tag{4.86}$$

when the boundary conditions at both horizontal walls are of no-slip, and when

$$\text{Ra} < \text{Ra}_{\text{crit.},2} = 657.511 \tag{4.87}$$

for free-slip. For higher values of the Rayleigh number, convection sets in. Remarkably, the values of $\text{Ra}_{crit.}$ for the initial transition is independent of the Prandtl number (which only plays a role after convective motion emerges).

---

[13] In athermal flows, there is an additional degree of freedom, because we can actually use a different fluid (i.e. change the viscosity) in each setup. This fact is often used in experimental fluid dynamics, to replace a large-scale flow system with one which fits within the scales of the laboratory or wind tunnel—as long as the dimensionless numbers are the same, the setups are equivalent in principle. For the RB setup, however, it is more difficult to exploit this degree of freedom, because of the requirement for having the same Pr value.

4.3.3 Numerical Algorithm Using the Lattice
Boltzmann Method (LBM)

To obtain numerical solutions of Eqs. (4.73)–(4.75), we use the *lattice Boltzmann method* (LBM)—a relatively new approach for solving numerically the fluid dynamics equations, based on simplified models inspired by statistical physics. Of the multitude of such models proposed in the literature, our implementation here will be based on the *multiple-relaxation-times* (MRT)[14] LBM approach, as described in Wang et al. [31]. We only provide a brief overview of the algorithms below. The reader interested in the topic can find more information in review articles such as Chen and Doolen [6] or Yu et al. [33], and in textbooks (in chronological order: Wolf-Gladrow [32], Succi [28], Sukop and Thorne [29], Mohamad [18], and Guo and Shu [13]). Chopard and Droz [7] discuss *cellular automata* (CA) and *lattice gas cellular automata* (LGCA), which are the precursors of the LBM.

Whereas in classical numerical analysis we start by discretizing the governing equations, a different approach is taken in LBM; here, the starting point is a discrete particle ("mesoscopic") model, which is then analyzed, to demonstrate that it recovers the equations of interest in the appropriate limits of vanishing space and time steps. Also resulting from this analysis are relations between some free parameters in the mesoscopic model and the physical parameters of the recovered equations.

LBM models are typically defined on a spatially-isotropic mesh ("lattice"). Each node of the mesh is populated by a number of fictitious particles, and each such particle has an associated *discretized velocity*. The quantities whose evolution is solved numerically by the algorithms are the *particle distribution functions* (PDFs) for each particle species, commonly denoted by f_i. For the nodes not near a domain boundary, the evolution of the (PDFs) at each discrete time step consists of a collision step (local to the node), followed by a streaming step (which is simply a shift in space of the PDF to the nearest neighbour specified by its associated discretized velocity). In mathematical notation, this reads:

$$f_i(\boldsymbol{x} + \boldsymbol{c}_i \delta_t, t + \delta_t) = f_i(\boldsymbol{x}, t) + \Omega\left(\boldsymbol{f}(\boldsymbol{x}, t)\right) \tag{4.88}$$

where $\boldsymbol{c}_i$ is the discretized velocity associated with particle species i, and Ω is the collision operator. Remarkably, solutions to various macroscopic equations can be recovered numerically by choosing suitable sets of discretized velocities and the collision operators. The solver we summarize below (following [31]) provides two examples of such choices, as the fluid and temperature equations are solved on two separate lattices.

---

[14] Even more precisely, the *two-relaxation-times* (TRT) subset of the MRT family is used, as some parameters are fixed.

Notation Conventions

In the presentation of the model below we use the following conventions:

- Prescripts $^\mathbb{F}\square$ or $^\mathbb{T}\square$ indicate to indicate that we refer to the fluid or temperature solvers respectively, when confusion may occur.
- Compared to the heat diffusion example from Sect. 4.1, we introduce an additional system of units (the *numerical* system; here—LBM). We will discuss in Sect. 4.3.4 how this is related to the dimensionless system of units. The superscript $\square^{(n)}$ denotes quantities in the numerical system.
- Finally, superscripts $\square^{\dagger}$ and $\square^{-1}$ denote the matrix transpose and inverse operations, respectively.

4.3.3.1 Model for the Fluid Component

The evolution equations for the fluid PDFs read:

$$\underbrace{f_i(\boldsymbol{x}^{(n)} +^{\mathbb{F}} \boldsymbol{c}_i \delta_t^{(n)}, \ t^{(n)} + \delta_t^{(n)}) = f_i(\boldsymbol{x}^{(n)}, \ t^{(n)})}_{\text{stream}} \underbrace{- M_{\alpha\beta}^{-1} \underbrace{S_{\beta\gamma}[\boldsymbol{m}_\gamma - \boldsymbol{m}_\gamma^{\text{eq}}]}_{\text{relax moments}}}_{\text{collide}} \quad (4.89)$$

where $\{i, \alpha, \beta, \gamma\} \in \{0, \ldots, 8\}$ and repeated Greek subscripts imply again summation.

There are 9 discretized velocities[15] for this model:

$$\left(^{\mathbb{F}}\boldsymbol{c}_0, \ldots, {}^{\mathbb{F}}\boldsymbol{c}_8\right) = c^{(n)} \begin{pmatrix} 0 & 1 & 0 & -1 & 0 & 1 & -1 & -1 & 1 \\ 0 & 0 & 1 & 0 & -1 & 1 & 1 & -1 & -1 \end{pmatrix} \quad (4.90)$$

where the basic lattice speed $c^{(n)}$—the same for the fluid and temperature solvers—is defined in terms of the discrete lattice spacing $\delta_x^{(n)}$ and time step $\delta_t^{(n)}$:

$$c^{(n)} \equiv \frac{\delta_x^{(n)}}{\delta_t^{(n)}} \quad (4.91)$$

For simplicity, it is customary to take $\delta_x^{(n)} = 1$ and $\delta_t^{(n)} = 1$.

The vector $\boldsymbol{m}$ consists of *moments*, which are evaluated (locally) from the PDFs through the linear transformation[16] specified by the matrix $\tilde{M}$:

---

[15] This particular topology is known as $D2Q9$ in the literature. The generic notation is $DdQq$, where d stands for the dimensionality of the lattice, and q for the number of particle species.

[16] The rows of $\tilde{M}$ represent orthogonal polynomials of the discretized velocities (see, e.g. Bouzidi et al. [2] or Dellar [10]).

$$m = \tilde{M} f \tag{4.92}$$

For the specific fluid solver chosen here, the transformation matrix reads:

$$\tilde{M} = \begin{pmatrix} 1 & 1 & 1 & 1 & 1 & 1 & 1 & 1 & 1 \\ 0 & 1 & 0 & -1 & 0 & 1 & -1 & -1 & 1 \\ 0 & 0 & 1 & 0 & -1 & 1 & 1 & -1 & -1 \\ -4 & -1 & -1 & -1 & -1 & 2 & 2 & 2 & 2 \\ 0 & 1 & -1 & 1 & -1 & 0 & 0 & 0 & 0 \\ 0 & 0 & 0 & 0 & 0 & 1 & -1 & 1 & -1 \\ 0 & -2 & 0 & 2 & 0 & 1 & -1 & -1 & 1 \\ 0 & 0 & -2 & 0 & 2 & 1 & 1 & -1 & -1 \\ 4 & -2 & -2 & -2 & -2 & 1 & 1 & 1 & 1 \end{pmatrix} \tag{4.93}$$

It can be verified that $\tilde{M} \cdot \tilde{M}^{\dagger}$ is a diagonal matrix, which simplifies the procedure for computing the inverse mapping (from moments to PDFs):

$$\tilde{M}^{-1} = \tilde{M}^{\dagger} \cdot \left(\tilde{M} \cdot \tilde{M}^{\dagger} \right)^{-1} \tag{4.94}$$

For the specific fluid model considered here, the components of m are:

$$m = (\rho^{(n)}, j_x^{(n)}, j_y^{(n)}, e^{(n)}, p_{xx}^{(n)}, p_{xy}^{(n)}, q_x^{(n)}, q_y^{(n)}, \epsilon^{(n)})^{\dagger} \tag{4.95}$$

which correspond to:

- $\rho^{(n)}$—fluid density
- $j_x^{(n)}, j_y^{(n)}$—x- and y-components of the fluid momentum
- $e^{(n)}$—fluid energy
- $p_{xx}^{(n)}, p_{xy}^{(n)}$—diagonal and off-diagonal components of the symmetric traceless viscous stress tensor
- $q_x^{(n)}, q_y^{(n)}$—x- and y-components of the energy flux
- $\epsilon^{(n)}$—related to square of the fluid energy

The equilibrium moments m^{eq} are:

$$m_0^{eq} = \delta \rho^{(n)} \tag{4.96}$$

$$m_1^{eq} = \rho_0^{(n)} u_x^{(n)} \tag{4.97}$$

$$m_2^{eq} = \rho_0^{(n)} u_y^{(n)} \tag{4.98}$$

$$m_3^{eq} = -2\delta \rho^{(n)} + 3\rho_0^{(n)} \left[(u_x^{(n)})^2 + (u_y^{(n)})^2 \right] \tag{4.99}$$

$$m_4^{eq} = \rho_0^{(n)} \left[(u_x^{(n)})^2 - (u_y^{(n)})^2 \right] \tag{4.100}$$

$$m_5^{eq} = \rho_0^{(n)} u_x^{(n)} u_y^{(n)} \tag{4.101}$$

$$m_6^{eq} = -\rho_0^{(n)} u_x^{(n)} \tag{4.102}$$

$$m_7^{eq} = -\rho_0^{(n)} u_y^{(n)} \tag{4.103}$$

$$m_8^{eq} = \delta\rho^{(n)} - 3\rho_0^{(n)} \left[(u_x^{(n)})^2 + (u_y^{(n)})^2 \right] \tag{4.104}$$

where the macroscopic variables are evaluated from the local PDFs:

$$\rho^{(n)} = \rho_0^{(n)} + \delta\rho^{(n)} \equiv \rho_0^{(n)} + \sum_{i=0}^{8} f_i \tag{4.105}$$

$$u_\gamma^{(n)} = \frac{1}{\rho_0^{(n)}} \sum_{i=0}^{8} \mathbb{F}_{c_{i,\gamma}} f_i, \quad \text{with} \quad \gamma \in \{x, y\} \tag{4.106}$$

where for convenience we choose $\rho_0^{(n)} = 1$ (reference density in the numerical system of units).

Finally, the relaxation matrix $\tilde{S}$ is a diagonal matrix:

$$\tilde{S} = \text{diag}(0, \ 1, \ 1, \ s_e, \ s_\nu, \ s_\nu, \ s_q, \ s_q, \ s_\epsilon) \tag{4.107}$$

There is some freedom as to how to choose these parameters, to optimize the stability of the model. For the incompressible *two-relaxation-times* (TRT) model we adopt here, $s_\nu = s_e = s_\epsilon$, and the matrix becomes:

$$\tilde{S} = \text{diag}(0, \ 1, \ 1, \ s_\nu, \ s_\nu, \ s_\nu, \ s_q, \ s_q, \ s_\nu) \tag{4.108}$$

where the adjustable parameter s_ν determines the kinematic viscosity of the model:

$$\nu^{(n)} = \frac{1}{3} \left(\frac{1}{s_\nu} - \frac{1}{2} \right) \tag{4.109}$$

and s_q is related to s_ν via:

$$s_q = 8\frac{2 - s_\nu}{8 - s_\nu} \tag{4.110}$$

For physical reasons, we need s_ν and $s_q \in [0, 2)$ (see Wang et al. [31] and Ginzburg and d'Humieres [12]).

Body forces: Strictly speaking, the evolution Eq. (4.89) only applies when there is no body force acting on the fluid. For simulating convective flows, as intended here, it would normally be necessary to extend this equation, by adding some correction terms to the RHS (e.g. as discussed by Guo et al. [14]). However, with the LBM-MRT class of models, the force can be added in a more natural way, directly to the corresponding moment (in our case—m_2 because the gravitational acceleration is along the y-coordinate). To recover the Navier-Stokes equations with a body force with 2nd-order accuracy, the force contribution to this moment is added in two stages, before and after the collision. This procedure is known as "Strang splitting" (see Dellar [11] and references therein for details).

boundary conditions (**BCs**): Our setup demands two types of BCs for the fluid solver:

- *periodic*: these can be easily enforced for our simple geometry directly at the implementation level, by constraining the streaming of PDFs along the y-direction using a modulo operation.
- *no-slip*: these are traditionally implemented in LBM via a procedure known as "bounce-back". In this approach, each post-collision PDF that would be moved to a solid node by normal streaming is instead copied to the local node (where it originated from a collision operation), but with the opposite orientation. This can be expressed mathematically as:

$$f_{\bar{i}}^{\text{pre-collision}}(x_f^{(n)}, t^{(n)} + \delta_t^{(n)}) = f_i^{\text{post-collision}}(x_f^{(n)}, t^{(n)}) \qquad (4.111)$$

where the overline is used to denote the discrete vector with opposite orientation:

$$-^{\text{F}} c_i = ^{\text{F}} c_{\bar{i}} \qquad (4.112)$$

and $x_f^{(n)}$ is the position of the fluid node adjacent to the solid boundary. This simple procedure effectively realizes a no-slip wall halfway between the fluid node and the neighbouring wall node.[17] Also, since no PDFs are "lost" or "gained", the approach conserves the total mass in the system.

4.3.3.2 Model for the Temperature Component

The model for solving the temperature advection-diffusion Eq. (4.53) is very similar to the one above corresponding to the fluid equations. However, a lattice with lower connectivity[18] is sufficient, because the temperature equations do not involve higher-order quantities like the stress tensor (which appears in the fluid equations). The evolution equations for the temperature PDFs read:

$$\underbrace{g_i(x^{(n)} + ^{\text{T}} c_i \delta_t^{(n)}, t^{(n)} + \delta_t^{(n)})}_{\text{stream}} = g_i(x^{(n)}, t^{(n)}) - \underbrace{N_{\alpha\beta}^{-1} Q_{\beta\gamma}[n_\gamma - n_\gamma^{\text{eq}}]}_{\text{relax moments}} \qquad (4.113)$$

$$\underbrace{\phantom{g_i(x^{(n)} + ^{\text{T}} c_i \delta_t^{(n)}, t^{(n)} + \delta_t^{(n)}) = g_i(x^{(n)}, t^{(n)}) - N_{\alpha\beta}^{-1} Q_{\beta\gamma}[n_\gamma - n_\gamma^{\text{eq}}]}}_{\text{collide}}$$

---

[17] This displacement of the boundary relative to the last fluid node needs to be taken into account during the initialization and postprocessing stages.
[18] Specifically, we use a $D2Q5$ lattice for temperature, while at least $D2Q9$ was necessary for the fluid.

where $\{i, \alpha, \beta, \gamma\} \in \{0, \ldots, 4\}$ and repeated Greek subscripts imply summation over the fictitious temperature particles.

The 5 discretized velocities for the model are:

$$\left({}^{\mathbb{T}}c_0, \ldots, {}^{\mathbb{T}}c_4\right) = c^{(n)} \begin{pmatrix} 0 & 1 & 0 & -1 & 0 \\ 0 & 0 & 1 & 0 & -1 \end{pmatrix} \tag{4.114}$$

where we use the same conventions as for the fluid solver above (i.e. $c^{(n)} = \delta_x^{(n)}/\delta_t^{(n)}$, with $\delta_x^{(n)} = 1$ and $\delta_t^{(n)} = 1$ for simplicity).

The local vector of moments $\boldsymbol{n}$ is recovered from the temperature PDFs through the linear transformation $\tilde{N}$:

$$\boldsymbol{n} = \tilde{N}\boldsymbol{g} \tag{4.115}$$

where:

$$\tilde{N} = \begin{pmatrix} 1 & 1 & 1 & 1 & 1 \\ 0 & 1 & 0 & -1 & 0 \\ 0 & 0 & 1 & 0 & -1 \\ -4 & 1 & 1 & 1 & 1 \\ 0 & 1 & -1 & 1 & -1 \end{pmatrix} \tag{4.116}$$

As for the analogue fluid matrix $\tilde{M}$, the product $\tilde{N} \cdot \tilde{N}^\dagger$ is a diagonal matrix, which simplifies the calculation of its inverse.

The equilibrium moments for temperature are defined as:

$$\boldsymbol{n}^{\mathrm{eq}} = (\theta^{(n)}, u_x^{(n)}\theta^{(n)}, u_y^{(n)}\theta^{(n)}, a\theta^{(n)}, 0)^\dagger \tag{4.117}$$

where:

- the macroscopic temperature is evaluated from the PDFs:

$$\theta^{(n)} = \sum_{i=0}^{4} g_i \tag{4.118}$$

- a is a model parameter which influences the thermal diffusivity
- $u_x^{(n)}$, $u_y^{(n)}$ represent the local components of the velocity, as evaluated from the fluid solver

The relaxation matrix $\tilde{Q}$ is again diagonal:

$$\tilde{Q} = \mathrm{diag}(0, \ \sigma_\kappa, \ \sigma_\kappa, \ \sigma_e, \ \sigma_\nu) \tag{4.119}$$

The parameter σ_κ, together with the parameter a from Eq. (4.117), determine the thermal diffusivity of the model:

$$\kappa^{(n)} = \frac{4+a}{10} \left(\frac{1}{\sigma_\kappa} - \frac{1}{2} \right) \tag{4.120}$$

As for the fluid model, stability and accuracy considerations restrict the possible values of the parameters σ_i and a. One particular choice, suitable for flows where the

$$\tilde{Q} = \mathrm{diag}(0,\ \sigma_\kappa,\ \sigma_\kappa,\ \sigma_\nu,\ \sigma_\nu) \tag{4.121}$$

where for reasons of accuracy and stability (see Wang et al. [31] for details) the relaxation matrix is fixed:

- $\sigma_\kappa = 3 - \sqrt{3}$
- $\sigma_\nu = 2(2\sqrt{3} - 3)$

and the thermal diffusivity of the model is instead determined by the parameter a:

$$\kappa^{(n)} = \frac{\sqrt{3}}{60}(4+a), \quad \text{with} \quad -4 < a < 1 \tag{4.122}$$

boundary conditions (**BCs**): Our setup requires two types of BCs for the temperature solver:

- *periodic*: this BC, necessary for the vertical walls, is again achieved at the implementation level, by "folding" the y-direction for the streaming step.
- *constant temperature*: several methods exist for imposing a constant temperature at the walls. One difficulty [16] is that maintaining a constant temperature also requires no heat conduction along the boundaries. To satisfy this condition with *2nd*-order accuracy, we use the same scheme as Wang et al. [31], consisting of a procedure known as "anti-bounce-back" procedure. Mathematically, this reads:

$$g_{\bar{i}}^{\text{pre-collision}}(x_f^{(n)}, t^{(n)} + \delta_t) = -g_i^{\text{post-collision}}(x_f^{(n)}, t^{(n)}) + 2\sqrt{3}\kappa^{(n)}\theta_{\text{wall}}^{(n)} \tag{4.123}$$

with the same meaning for the overline (opposing direction).

4.3.4 Connecting the Numerical and Dimensionless Systems of Units

While in Sect. 4.1.2 we discretized our heat diffusion problem directly in the dimensionless system of units, for the current problem we work with yet another system of units—the *numerical* system. This is beneficial here, since the LBM algorithm requires several constraints on the parameters to hold, as discussed earlier. However, these are not connected (at least not in an obvious manner) to the actual physics in the system, and it helps to draw a distinction between the system in which computations are actually performed (where the characteristics of the algorithm are the main concern) and the dimensionless system, where the physical setup is emphasized.

The mapping between the two systems of units is very similar to the one discussed above, for non-dimensionalizing the physical equations. We choose N_y (number of nodes along the channel's height) as the spatial scale and N_t (number of iterations to represent one characteristic time interval—to be specified later) as the temporal scale. Remembering, also, that the methods we use for enforcing the BCs at the horizontal walls place the effective boundary halfway between the fluid and solid walls, we can choose the following scaling relations:

$$x_\beta^{(d)} = \frac{x_\beta^{(n)} - \frac{1}{2}}{N_y} \Leftrightarrow x_\beta^{(n)} = \frac{1}{2} + N_y x_\beta^{(d)} \qquad (4.124)$$

$$\partial_\beta^{(d)} = N_y \partial_\beta^{(n)} \Leftrightarrow \partial_\beta^{(n)} = \frac{1}{N_y} \partial_\beta^{(d)} \qquad (4.125)$$

$$\partial_{\beta\gamma}^{(d)} = N_y^2 \partial_{\beta\gamma}^{(n)} \Longleftrightarrow \partial_{\beta\gamma}^{(n)} = \frac{1}{N_y^2} \partial_{\beta\gamma}^{(d)} \qquad (4.126)$$

$$t^{(d)} = \frac{1}{N_t} t^{(n)} \Longleftrightarrow t^{(n)} = N_t t^{(d)} \qquad (4.127)$$

$$\partial_t^{(d)} = N_t \partial_t^{(n)} \Longleftrightarrow \partial_t^{(n)} = \frac{1}{N_t} \partial_t^{(d)} \qquad (4.128)$$

$$u_\beta^{(d)} = \frac{N_t}{N_y} u_\beta^{(n)} \Longleftrightarrow u_\beta^{(n)} = \frac{N_y}{N_t} u_\beta^{(d)} \qquad (4.129)$$

$$p^{(d)} = \frac{1}{3} \left(\frac{N_t}{N_y} \right)^2 \delta\rho^{(n)} \Longleftrightarrow \delta\rho^{(n)} = 3 \left(\frac{N_y}{N_t} \right)^2 p^{(d)} \qquad (4.130)$$

$$\theta^{(d)} = \theta^{(n)} \qquad (4.131)$$

where we used the equation of state of the LBM solver:

$$p^{(n)} = \frac{1}{3} (\delta\rho)^{(n)} \qquad (4.132)$$

to translate directly between dimensionless pressure and the solver's numerical density anomalies.

Plugging the equations above into the dimensionless governing equations, we obtain the following expressions for the model parameters:

$$\nu^{(n)} = \frac{\mathrm{Pr}\, N_y^2}{N_t} \qquad (4.133)$$

$$(\alpha g)^{(n)} = \frac{\mathrm{Ra}\,\mathrm{Pr}\, N_y}{N_t^2} \qquad (4.134)$$

$$\kappa^{(n)} = \frac{N_y^2}{N_t} \tag{4.135}$$

To complete the formulation of the problem in the numerical system, we have the following BCs for temperature:

$$\theta\left(x^{(n)}, \frac{1}{2}, t^{(n)}\right) = +\frac{1}{2} \tag{4.136}$$

$$\theta\left(x^{(n)}, N_y + \frac{1}{2}, t^{(n)}\right) = -\frac{1}{2}, \tag{4.137}$$

and the following initial profiles for temperature and density anomaly:

$$\theta^{(n)}(x^{(n)}, y^{(n)}, 0) = \frac{1}{2} - \frac{1}{2N_y}(2y^{(n)} - 1) \tag{4.138}$$

$$(\delta\rho)^{(n)}(x^{(n)}, y^{(n)}, 0) = \frac{3(\alpha g)^{(n)}}{2N_y}\left(y^{(n)} - \frac{1}{2}\right)\left(N_y + \frac{1}{2} - y^{(n)}\right) \tag{4.139}$$

4.3.5 Numerical Implementation in Fortran (OOP)

As for the heat diffusion solver described in Sect. 4.1, we construct our implementation around the OOP methodology. However, we organize the solution differently here, because of the increased complexity of the numerical algorithm, and to illustrate some additional techniques. As for the other case studies, we describe the main entities below (see file $\boxed{\texttt{lbm2d\_mrt\_rb\_v1.f90}}$ for the complete code).

module *NumericKinds*: Even more than in our previous example application (Sect. 4.2), the range and accuracy of the numeric types used in our fluid solver is crucial. Therefore, we use the same mechanisms as before, to allow convenient and reliable selection of the precision of the variables. As a small enhancement of this module for this application, we provide appropriate'swap'-subroutines, grouped under a generic interface:

```fortran
module NumericKinds
  implicit none

  ! KINDs for different types of REALs
  ! .................................
  integer, parameter :: RK = R_DP ! if changing this, also change RK_FMT

  ! KINDs for different types of INTEGERs
  ! .....................................
  integer, parameter :: IK = I3B

  ! Edit-descriptors for real-values
  character(len=*), parameter :: R_SP_FMT = "f0.6", &
      R_DP_FMT = "f0.15", R_QP_FMT = "f0.33"
  ! Alias for output-precision to use in the program (keep this in sync with RK)
  character(len=*), parameter :: RK_FMT = R_DP_FMT

  interface swap ! generic IFACE
      module procedure swapRealRK, swapIntIK
```

```
    end interface swap
contains

    elemental subroutine swapRealRK( a, b )
       real(RK), intent(inout) :: a, b
       real(RK) :: tmp
       tmp = a; a = b; b = tmp
    end subroutine swapRealRK

    elemental subroutine swapIntIK( a, b )
       integer(IK), intent(inout) :: a, b
       integer(IK) :: tmp
       tmp = a; a = b; b = tmp
    end subroutine swapIntIK
end module NumericKinds
```

Listing 4.10 `src/Chapter4/lbm2d_mrt_rb_v1.f90` (excerpt)

modules `LbmConstantsMrtD2Q5` **and** `LbmConstantsMrtD2Q9` : These
two additional modules (not reproduced here for brevity) are used for specifying fixed
model parameters for the LBM solver. For example, we have here the directions of the
discretized velocities and the matrix operators which map distributions to moments
(and the other way around).

`MrtSolverBoussinesq2D` **type**: The core of the numerical solver is implemented as procedures bound to this type. Since the numerical method is, in principle,
not restricted to a particular setup,[19] we avoid including here any constants specific
to the RB problem. The solver class (outlined below) is supported by type-bound
procedures similar to those of the heat diffusion solver from Sect. 4.1, except that the
task of writing output is delegated to other classes (explained below):

```
module MrtSolverBoussinesq2D_class
   use NumericKinds, only : IK, RK, swap
   use LbmConstantsMrtD2Q5
   use LbmConstantsMrtD2Q9
   implicit none

   type :: MrtSolverBoussinesq2D
      private
      ! parameters for the algorithm (not bound to the RB-setup)
      real(RK) :: mAlphaG, mAParam, mViscosity, mDiffusivity, &
         mTempColdWall, mTempHotWall, &
         ! for relaxation-matrices, we store only the non-zero part (= diagonals)
         mRelaxVecFluid(0:8), mRelaxVecTemp(0:4)

      ! internal model arrays
      ! NOTES: - last dimension is for 2-lattice alternation
      !        - 1st dimension: 0-9 = fluid, 9-13 = temp DFs
      real(RK), dimension(:,:,:,:), allocatable :: mDFs
      ! raw moments from which we can compute macroscopic fields; this is used
      ! mainly for simulation-output
      ! 0 - pressure| 1 - uX| 2 - uY| 3 - temp
      real(RK), dimension(:,:,:), allocatable :: mRawMacros

      integer(IK) :: mOld, mNew, & ! for tracking most recent lattice
         mNx, mNy ! mesh-size (received from 'sim'-class)

   contains
      private
      procedure, public :: init => initMrtSolverBoussinesq2D
      procedure, public :: advanceTime => advanceTimeMrtSolverBoussinesq2D
      procedure, public :: cleanup => cleanupMrtSolverBoussinesq2D
      procedure, public :: getRawMacros => getRawMacrosMrtSolverBoussinesq2D
      ! internal methods
```

---

[19] Note, however, that the boundary conditions are hard-coded into the solver, for simplicity. To simulate a problem with different BCs, it is necessary to modify the procedure
`advanceTimeMrtSolverBoussinesq2D` (which implements the actual LBM-dynamics).
The interested reader may remove this hard-coding by adding "mask"-arrays, which classify the
different types of nodes (e.g. bulk, no-slip, etc. for the fluid component and bulk, constant temperature, and adiabatic for the temperature component). Also, the code for enforcing these different
types of BCs can be further isolated into distinct procedures, or even into different classes (useful
for BC-algorithms which also need to hold some own data, e.g. the temperature at the boundary).

```fortran
        procedure :: calcLocalMomsMrtSolverBoussinesq2D
        procedure :: calcLocalEqMomsMrtSolverBoussinesq2D
    end type MrtSolverBoussinesq2D

contains
    function getRawMacrosMrtSolverBoussinesq2D( this ) result(macros)
        class(MrtSolverBoussinesq2D), intent(in) :: this
        real(RK), dimension(this%mNx, this%mNy, 0:3) :: macros
        macros = this%mRawMacros
    end function getRawMacrosMrtSolverBoussinesq2D

    subroutine initMrtSolverBoussinesq2D( this, nX, nY, tempColdWall, tempHotWall, &
            viscosity, diffusivity, alphaG, aParam, relaxVecFluid, relaxVecTemp )
        class(MrtSolverBoussinesq2D), intent(inout) :: this
        real(RK), intent(in) :: tempColdWall, tempHotWall, &
            viscosity, diffusivity, alphaG, aParam, &
            relaxVecFluid(0:8), relaxVecTemp(0:4)
        integer(IK), intent(in) :: nX, nY
        integer(IK) :: x, y, i ! dummy vars
        integer(IK), dimension(0:1) :: dest
        ! temporary moments-vars
        real(RK) :: fluidMoms(0:8), tempMoms(0:4), tempPerturbation

        ! copy argument-values internally
        this%mNx = nX; this%mNy = nY
        this%mTempColdWall = tempColdWall; this%mTempHotWall = tempHotWall
        this%mViscosity = viscosity; this%mDiffusivity = diffusivity
        this%mAlphaG = alphaG; this%mAParam = aParam
        this%mRelaxVecFluid = relaxVecFluid; this%mRelaxVecTemp = relaxVecTemp

        tempPerturbation=this%mTempHotWall/1.E5_RK

        ! get memory for model-state arrays (and Y-buffers)
        allocate( this%mDFs(0:13, 1:this%mNx, 0:(this%mNy+1), 0:1) )
        allocate( this%mRawMacros(this%mNx, this%mNy, 0:3) )

        ! initialize
        this%mDFs = 0._RK
        this%mRawMacros = 0._RK

        ! init tracking-vars for lattice-alternation
        this%mOld = 0; this%mNew = 1

        ! ICs for model's state-arrays
        do y=1, this%mNy
            do x=1, this%mNx
                ! reset moments-vectors
                fluidMoms = 0._RK; tempMoms = 0._RK
                ! Initialize pressure with steady-state (quadratic) profile, to avoid
                ! the initial oscillations.
                fluidMoms(0) = &
                    (3._RK*this%mAlphaG)/(2._RK*this%mNy)*(y-0.5_RK)*(this%mNy+0.5_RK-y)

                ! Initialize temperature with steady-state (linear) profile, to save
                ! CPU-time. Also here, we insert small perturbation, to break the
                ! symmetry of the system (otherwise, the simulation is too stable).
                tempMoms(0) = 0.5_RK - (2._RK*y-1._RK)/(2._RK*this%mNy)
                if( (x == this%mNx/3+1) .and. (y == 2) ) then
                    tempMoms(0) = tempMoms(0)+tempPerturbation
                end if

                ! map moments onto DFs...
                ! ...fluid
                do i=0, 8
                    this%mDFs(i, x, y, this%mOld) = dot_product(M_INV_FLUID(:,i), fluidMoms)
                end do
                ! ...temp
                do i=0, 4
                    this%mDFs(i+9, x, y, this%mOld) = dot_product(N_INV_TEMP(:,i), tempMoms)
                end do

                ! Fill buffers for bounce-back (for initial time-step)
                ! ...fluid
                do i=0, 8
                    dest(0) = mod(x+EV_FLUID(1, i)+this%mNx-1, this%mNx)+1
                    dest(1) = y+EV_FLUID(2, i)
                    if( (dest(1) == 0) .or. (dest(1) == this%mNy+1) ) then
                        this%mDFs(i, dest(0), dest(1), this%mOld) = &
                            this%mDFs(i, x, y, this%mOld)
                    end if
                end do

                ! ...temp
                do i=0, 4
                    dest(0) = mod(x+EV_TEMP(1, i)+this%mNx-1, this%mNx)+1
                    dest(1) = y+EV_TEMP(2, i)
                    if( (dest(1) == 0) .or. (dest(1) == this%mNy+1) ) then
                        this%mDFs(i+9, dest(0), dest(1), this%mOld) = &
                            this%mDFs(i+9, x, y, this%mOld)
                    end if
                end do

                ! save ICs
                this%mRawMacros(x, y, :) = [ fluidMoms(0:2), tempMoms(0) ]
            end do
        end do
    end subroutine initMrtSolverBoussinesq2D

    subroutine calcLocalMomsMrtSolverBoussinesq2D( this, x, y, fluidMoms, tempMoms )
        class(MrtSolverBoussinesq2D), intent(in) :: this
        integer(IK), intent(in) :: x, y
```

```fortran
      real(RK), intent(out) :: fluidMoms(0:8), tempMoms(0:4)
      ! ..........................
  end subroutine calcLocalMomsMrtSolverBoussinesq2D

  subroutine calcLocalEqMomsMrtSolverBoussinesq2D( this, &
        dRho, uX, uY, temp, fluidEqMoms, tempEqMoms )
      class(MrtSolverBoussinesq2D), intent(inout) :: this
      real(RK), intent(in) :: dRho, uX, uY, temp
      real(RK), intent(out) :: fluidEqMoms(0:8), tempEqMoms(0:4)
      ! ..........................
  end subroutine calcLocalEqMomsMrtSolverBoussinesq2D

  ! advance solver-state by one time-step (core LBM-algorithm)
  subroutine advanceTimeMrtSolverBoussinesq2D( this )
      class(MrtSolverBoussinesq2D), intent(inout) :: this
      ! local vars
      integer(IK) :: x, y, i, old, new ! dummy indices
      integer(IK), dimension(0:1) :: dest
      real(RK) :: fluidMoms(0:8), tempMoms(0:4), &
          fluidEqMoms(0:8), tempEqMoms(0:4)

      ! initializations
      dest = 0; fluidMoms = 0._RK; tempMoms = 0._RK
      fluidEqMoms = 0._RK; tempEqMoms = 0._RK
      old = this%mOld; new = this%mNew

      do y=1, this%mNy
        do x=1, this%mNx
          call this%calcLocalMomsMrtSolverBoussinesq2D(x, y, fluidMoms, tempMoms)

          ! add 1st-half of force term (Strang splitting)
          fluidMoms(2) = fluidMoms(2) + this%mAlphaG*0.5_RK*tempMoms(0)

          ! save moments related to output
          this%mRawMacros(x, y, :) = &
            [ fluidMoms(0), fluidMoms(1), fluidMoms(2), tempMoms(0) ]

          call this%calcLocalEqMomsMrtSolverBoussinesq2D( dRho=fluidMoms(0), &
              uX=fluidMoms(1), uY=fluidMoms(2), temp=tempMoms(0), &
              fluidEqMoms=fluidEqMoms, tempEqMoms=tempEqMoms )

          ! collision (in moment-space)
          fluidMoms = fluidMoms - this%mRelaxVecFluid * (fluidMoms - fluidEqMoms)
          tempMoms = tempMoms - this%mRelaxVecTemp * (tempMoms - tempEqMoms)

          ! add 2nd-half of force term (Strang splitting)
          fluidMoms(2) = fluidMoms(2) + this%mAlphaG*0.5_RK*tempMoms(0)

          ! map moments back onto DFs...
          ! ...fluid
          do i=0, 8
            this%mDFs(i, x, y, old) = dot_product( M_INV_FLUID(:, i), fluidMoms )
          end do
          ! ...temp
          do i=0, 4
            this%mDFs(i+9, x, y, old) = dot_product( N_INV_TEMP(:, i), tempMoms )
          end do

          ! stream to new array...
          ! ...fluid
          do i=0, 8
            dest(0) = mod(x+EV_FLUID(1, i)+this%mNx-1, this%mNx)+1
            dest(1) = y+EV_FLUID(2, i)
            ! STREAM (also storing runaway DFs in Y-buffer space)
            this%mDFs(i, dest(0), dest(1), new) = this%mDFs(i, x, y, old)
            if( dest(1) == 0 ) then
              if( EV_FLUID(2, i) /= 0 ) then
                ! apply bounce-back @bottom
                this%mDFs(OPPOSITE_FLUID(i), x, y, new) = &
                    this%mDFs(i, dest(0), dest(1), old)
              end if
            elseif( dest(1) == this%mNy+1 ) then
              if( EV_FLUID(2, i) /= 0 ) then
                ! apply bounce-back @top
                this%mDFs(OPPOSITE_FLUID(i), x, y, new) = &
                    this%mDFs(i, dest(0), dest(1), old)
              end if
            end if
          end do
          ! ...temp
          do i=0, 4
            dest(0) = mod(x+EV_TEMP(1, i)+this%mNx-1, this%mNx)+1
            dest(1) = y+EV_TEMP(2, i)
            ! STREAM (also storing runaway DFs in Y-buffer space)
            this%mDFs(i+9, dest(0), dest(1), new) = this%mDFs(i+9, x, y, old)
            if( dest(1) == 0 ) then
              ! apply anti-bounce-back @bottom
              this%mDFs(OPPOSITE_TEMP(i)+9, x, y, new) = &
                  -this%mDFs(i+9, dest(0), dest(1), old) + &
                  2._RK*sqrt(3._RK)*this%mDiffusivity*this%mTempHotWall
            elseif( dest(1) == this%mNy+1 ) then
              ! apply anti-bounce-back @top
              this%mDFs(OPPOSITE_TEMP(i)+9, x, y, new) = &
                  -this%mDFs(i+9, dest(0), dest(1), old) + &
                  2._RK*sqrt(3._RK)*this%mDiffusivity*this%mTempColdWall
            end if
          end do
        end do
      end do

      ! swap 'pointers' (for lattice-alternation)
```

```
      call swap( this%mOld, this%mNew )
   end subroutine advanceTimeMrtSolverBoussinesq2D

   subroutine cleanupMrtSolverBoussinesq2D( this )
      class(MrtSolverBoussinesq2D), intent(inout) :: this
      deallocate( this%mDFs, this%mRawMacros )  ! release memory
   end subroutine cleanupMrtSolverBoussinesq2D
end module MrtSolverBoussinesq2D_class
```

Listing 4.11 `src/Chapter4/lbm2d_mrt_rb_v1.f90` (excerpt)

`RBenardSimulation` **type**: To continue the comparison with the heat diffusion code, the role of the `Config` there is taken by the `RBenardSimulation` in our current case. However, some additional functionality is aggregated here, as the class is also orchestrating operations of the solver, and ensuring that computations and writing of output are properly synchronized:

```
module RBenardSimulation_class
   use NumericKinds, only : IK, RK
   use MrtSolverBoussinesq2D_class
   use OutputAscii_class
   implicit none

   ! Fixed simulation-parameters
   real(RK), parameter :: &
      ! To allow the 1st instability to develop, the aspect-ratio needs to be a
      ! multiple of 2π/kc, where kc = 3.117 (see [Shan1997]).
      ASPECT_RATIO = 2*2.0158, &
      ! See [Wang2013] for justification of these parameters.
      SIGMA_K = 3._RK - sqrt(3._RK), &
      SIGMA_NU_E = 2._RK * (2._RK*sqrt(3._RK) - 3._RK), &
      TEMP_COLD_WALL = -0.5, TEMP_HOT_WALL = +0.5

   type :: RBenardSimulation
      private
      integer(IK) :: mNx, mNy, & ! lattice size
         mNumIters1CharTime, mNumItersMax, &
         mNumOutSlices ! user-setting

      type(MrtSolverBoussinesq2D) :: mSolver ! associated solver...
      type(OutputAscii) :: mOutSink ! ...and output-writer

   contains
      private
      procedure, public :: init => initRBenardSimulation
      procedure, public :: run => runRBenardSimulation
      procedure, public :: cleanup => cleanupRBenardSimulation
   end type RBenardSimulation

contains
   subroutine initRBenardSimulation( this, Ra, Pr, nY, simTime, maxMach, &
         numOutSlices, outFilePrefix )
      class(RBenardSimulation), intent(out) :: this
      real(RK), intent(in) :: Ra, Pr, simTime, maxMach
      integer(IK), intent(in) :: nY, numOutSlices
      character(len=*), intent(in) :: outFilePrefix
      ! ..............................
   end subroutine initRBenardSimulation

   subroutine runRBenardSimulation( this )
      class(RBenardSimulation), intent(inout) :: this
      integer(IK) :: currIterNum ! dummy index
      real(RK) :: tic, toc ! for performance-reporting

      call cpu_time(time=tic) ! serial

      ! MAIN loop (time-iteration)
      do currIterNum=1, this%mNumItersMax
         ! simple progress-monitor
         if( mod(currIterNum-1, (this%mnumitersmax-1)/10) == 0 ) then
            write(*, '(i5,a)') nint((currIterNum*100._RK)/this%mnumitersmax), "%"
         end if

         call this%mSolver%advanceTime()

         call this%mOutSink%writeOutput( this%mSolver%getRawMacros(), currIterNum )
      end do

      call cpu_time(time=toc) ! serial

      write(*, '(/,a,f0.2,a)') "Performance Information: achieved", &
         this%mNumItersMax*real(this%mNx*this%mNy, RK) / (1.0e6*(toc-tic)), &
         "MLUPS (mega-lattice-updates-per-second)"
   end subroutine runRBenardSimulation

   subroutine cleanupRBenardSimulation( this )
      class(RBenardSimulation), intent(inout) :: this

      call this%mSolver%cleanup()
      call this%mOutSink%cleanup()
```

```
      end subroutine cleanupRBenardSimulation
  end module RBenardSimulation_class
```

Listing 4.12 `src/Chapter4/lbm2d_mrt_rb_v1.f90` (excerpt)

`OutputBase` **and** `OutputAscii` **types**: For this initial implementation, we write data on disk as simple ASCII files. However, this approach is far from ideal, because it forces many numeric-to-string conversions (which increase the output overhead), and also occupies more space on disk. Because of this, we only write the temperature field and maximum vertical velocity for now, and postpone writing of all simulation fields until Sect. 5.2.2, where we extend this application to demonstrate writing in the net CDF-format. To avoid duplication of code at that stage, we structure the implementation of the output functionality as a small hierarchy based on type extension (inheritance), whereby code that does not depend on the ultimate file format (e.g. some initializations of the conversion factors from the numerical to the dimensionless systems of units, axis coordinate, and implementation of the output criterion) is grouped under the `OutputBase` type:

```
module OutputBase_class
  use NumericKinds, only : IK, RK
  implicit none

  ! string-constants for output metadata
  character(len=*), parameter : : UNITS_STR="units", & ! for global-attribute
      SPACE_UNITS_STR="char. length", TIME_UNITS_STR="char. time", &
      PRESS_UNITS_STR="char. pressure-difference", &
      VEL_UNITS_STR="char. velocity", TEMP_UNITS_STR="char. temperature-difference"

  type : : OutputBase
    real(RK) : : mUyMax
    character(len=256) : : mOutFilePrefix

    ! information about the simulation
    integer(IK) : : mNx, mNy, mNumOutSlices, mNumItersMax, mOutDelay, mOutInterv
    real(RK) : : mDxD, mDtD, mRa, mPr, mMaxMach
    integer(IK) : : mCurrOutSlice ! for tracking output time-slices

    ! arrays for coordinates along each dimension (space & time)
    real(RK), dimension(:), allocatable : : mXVals, mYVals, &
        mTVals ! 1st output-slice - t=0 (ICs)

    ! conversion-factors for translating output from numerical- to
    ! dimensionless-units
    real(RK) : : mDRhoSolver2PressDimless, mVelSolver2VelDimless
  contains
    private
    procedure, public : : init => initOutputBase
    procedure, public : : cleanup => cleanupOutputBase
    procedure, public : : isActive => isActiveOutputBase
    procedure, public : : isTimeToWrite => isTimeToWriteOutputBase
  end type OutputBase
contains

  subroutine initOutputBase( this, nX, nY, numOutSlices, dxD, dtD, &
      nItersMax, outFilePrefix, Ra, Pr, maxMach )
    class(OutputBase), intent(inout) : : this
    integer(IK), intent(in) : : nX, nY, nItersMax, numOutSlices
    real(RK), intent(in) : : dxD, dtD, Ra, Pr, maxMach
    character(len=*), intent(in) : : outFilePrefix
    ! local vars
    integer(IK) : : x, y, t, tOut

    if( numOutSlices < 0 ) then
      this%mNumOutSlices = nItersMax + 1 ! write everything
    else
      this%mNumOutSlices = numOutSlices
    end if

    if( this%isActive() ) then ! prepare output only if actually writing
      ! copy over remaining arguments into internal-state
      this%mNx = nX; this%mNy = nY
      this%mNumItersMax = nItersMax
      this%mDxD = dxD; this%mDtD = dtD
      this%mRa = Ra; this%mPr = Pr; this%mMaxMach = maxMach
      this%mOutFilePrefix = outFilePrefix

      ! conversion-factors for output
      this%mVelSolver2VelDimless = dxD / dtD
      this%mDRhoSolver2PressDimless = this%mVelSolver2VelDimless**2 / 3._RK

      ! get memory for (dimensionless) coordinate-arrays
```

```
         allocate(this%mXVals(nX), this%mYVals(nY), this%mTVals(this%mNumOutSlices))

         ! Enforce safety-check: cannot request more output-slices than nItersMax!
         if( this%mNumOutSlices > (this%mNumItersMax+1) ) then
            write(*,'(3(a,/),a)') "ERROR: invalid combination of output-parameters", &
               "Cause: numOutSlices too high for computed number of iterations!", &
               "Fixes: decrease numOutSlices OR increase simTime", &
               "Aborting..."
            stop
         elseif( this%mNumOutSlices <= 0 ) then
         end if

         if( this%mNumOutSlices > 1 ) then ! avoid divide-by-zero if writing only ICs
            ! write output (mostly) every 'mOutInterv' iters...
            this%mOutInterv = this%mNumItersMax / (this%mNumOutSlices-1)
            ! ...except in the beginning
            this%mOutDelay = mod( this%mNumItersMax, this%mNumOutSlices-1 ) + 1
         else ! mNumOutSlices can only be 1
            this%mOutInterv = 1
         end if

         ! fill-in coordinate-arrays...
         ! ...X-dimension
         do x=1, this%mNx
            this%mXVals(x) = real( dxD*(x-0.5) )
         end do
         ! ...Y-dimension
         do y=1, this%mNy
            this%mYVals(y) = real( dxD*(y-0.5) )
         end do
         ! ...time-dimension
         ! NOTE: the time-delay between the first two output-slices is different from
         ! the subsequent ones, because:
         ! a) the ICs are written as 1st output-slice and
         ! b) the total number of model-iterations is not necessarily a multiple of
         ! the number of output-slices requested by the user
         this%mTVals(1) = real(0) ! ICs
         if( this%mNumOutSlices > 1 ) then
            tOut = this%mNumOutSlices
            do t=this%mNumItersMax, this%mOutDelay, -this%mOutInterv
               this%mTVals(tOut) = real( (t-0.5)*dtD )
               tOut = tOut-1 ! decrement time-slice index
            end do
         end if

         this%mCurrOutSlice = 0
      else
         write(*,'(a)') "INFO: no file-output, due to chosen 'numOutSlices'"
      end if
   end subroutine initOutputBase

   subroutine cleanupOutputBase( this )
      class(OutputBase), intent(inout) : : this
      if( this%isActive() ) then
         deallocate( this%mXVals, this%mYVals, this%mTVals )
      end if
   end subroutine cleanupOutputBase

   logical function isActiveOutputBase( this )
      class(OutputBase), intent(in) : : this
      isActiveOutputBase = ( this%mNumOutSlices > 0 )
   end function isActiveOutputBase

   ! Implement criterion for determining at which iterations to write output,
   ! based on the number of time-slices requested by user, subject to the
   ! constraints of:
   ! a) writing the ICs
   ! b) writing the last iteration (since we made the effort to compute so far in
   ! the first place)
   ! c) having equidistant (in time) output-slices (except for the initial delay)
   logical function isTimeToWriteOutputBase( this, iterNum )
      class(OutputBase), intent(in) : : this
      integer(IK), intent(in) : : iterNum

      if( this%isActive() ) then
         isTimeToWriteOutputBase = (iterNum==0) .or. ( &
            (this%mNumOutSlices /= 1) .and. &
            (iterNum >= this%mOutDelay) .and. &
            (mod(this%mNumItersMax-iterNum, this%mOutInterv) == 0) )
      else
         isTimeToWriteOutputBase = .false.
      end if
   end function isTimeToWriteOutputBase
end module OutputBase_class
```

Listing 4.13 | `src/Chapter4/lbm2d_mrt_rb_v1.f90` | (excerpt)

From this we derive the child types, which handle the peculiarities of each output
format. In this version of the program, the only child is `OutputAscii`:

```fortran
module OutputAscii_class
   use NumericKinds, only : IK, RK, RK_FMT
   use OutputBase_class
   implicit none

   type, extends(OutputBase) : : OutputAscii
   private
   integer(IK) : : mSummaryFileUnit, mTempFileUnit
   character(len=256) : : mSummaryFileName, mTempFileName, &
      mFmtStrngFieldFileNames
contains
   private
   ! public methods which differ from base-class analogues
   procedure, public : : init => initOutputAscii
   procedure, public : : writeOutput => writeOutputAscii
   procedure, public : : cleanup => cleanupOutputAscii
   ! internal method(s)
   procedure writeSummaryFileHeaderOutputAscii
end type OutputAscii contains subroutine initOutputAscii( this, nX,
nY, numOutSlices, dxD, dtD, &
   nItersMax, outFilePrefix, Ra, Pr, maxMach )
   class(OutputAscii), intent(inout) : : this
   integer(IK), intent(in) : : nX, nY, nItersMax, numOutSlices
   real(RK), intent(in) : : dxD, dtD, Ra, Pr, maxMach
   character(len=*), intent(in) : : outFilePrefix
   ! ...........................
   ! local

subroutine writeSummaryFileHeaderOutputAscii( this, fileUnit )
   class(OutputAscii), intent(inout) : : this
   integer(IK), intent(in) : : fileUnit
   ! ...........................
   ! local

subroutine writeOutputAscii( this, rawMacros, iterNum )
   class(OutputAscii), intent(inout) : : this
   real(RK), dimension(:, :, 0:), intent(in) : : rawMacros
   integer(IK), intent(in) : : iterNum
   ! ...........................
   ! local vars

subroutine cleanupOutputAscii( this )
   class(OutputAscii), intent(inout) : : this
   if( this%isActive() ) then
      write(this%mSummaryFileUnit, '(a)') " # UyMax"
      close(this%mSummaryFileUnit)
      call this%OutputBase%cleanup()
   end if
end subroutine cleanupOutputAscii
end module OutputAscii_class
```

Listing 4.14 `src/Chapter4/lbm2d_mrt_rb_v1.f90` (excerpt)

The main-program, shown below, is very compact:

```fortran
program lbm2d_mrt_rb
   use RBenardSimulation_class
   implicit none

   type(RBenardSimulation) : : testSim

   call testSim%init( Ra=1900._RK, Pr=7.1_RK, &
      nY=62, simTime=15._RK, maxMach=0.3_RK, &
      numOutSlices=80, outFilePrefix="rb_Ra_1900" )

   call testSim%run()

   call testSim%cleanup()
end program lbm2d_mrt_rb
```

Listing 4.15 `src/Chapter4/lbm2d_mrt_rb_v1.f90` (excerpt)

An instance of the simulation type is declared and then initialized. The arguments in the initialization call have the following meaning:

- $\boxed{\text{Ra}}$, $\boxed{\text{Pr}}$ —dimensionless numbers which determine the dynamics of the system

- **nY**—number of nodes used to simulate the channel's height; the width is then computed automatically, based on a pre-defined aspect ratio (declared in module `RBenardSimulation_class`)
- **simTime**—the total time to be simulated by the solver, in multiples of the reference time
- **maxMach**—this can be interpreted here as just another model parameter, which controls another source of model errors (compressibility error); it can be decreased, to improve the accuracy of the results (the corresponding error term is proportional to the square of this value)
- **numOutSlices**—number of time steps to appear in the output files

  1. *numOutSlices* < 0 causes all time steps to be written to disk
  2. *numOutSlices* = 0 suppresses output
  3. *numOutSlices* > 0 results in *numOutSlices* being written (including one time slice for the ICs of the simulation).

In Fig. 4.5 we present a sample temperature contour plot, for Ra = 1,900 and Pr = 7.1 (when the flow is already unstable). The plot was produced with the R-scrips `plotFieldFromAscii.R` (also available in the source code repository).

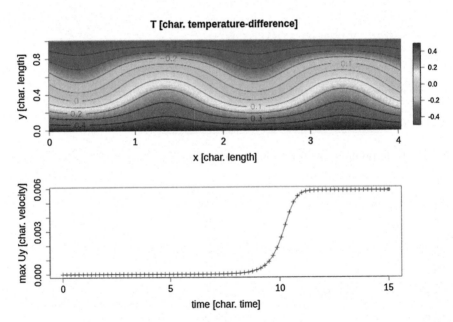

Fig. 4.5 Sample numerical solution for the RB problem, for Ra = 1,900 and Pr = 7.1; the upper plot is a visualization of the temperature profile at $t^{(d)} = 15$, while the lower plot shows the evolution of the maximum vertical velocity in the simulation over the period $t^{(d)} \in [0, 15]$

Exercise 18 (*Critical Rayleigh number* (Ra)) Using the program described in this section, perform several numerical experiments, to validate the critical value of the Rayleigh number, for which convection first develops. In each experiment, select a different value for the Rayleigh number, while keeping the other simulation parameters constant.

Hints:

- The current version of the program (`lbm2d_mrt_rb_v1.f90`) is not very efficient, so we recommend to perform only two runs (for example, Ra = 1,500 and Ra = 2,000, to limit the computational effort. In Chap. 5 we discuss far more efficient versions of this application.

- For estimating the critical Ra, it is convenient to study the growth rate of the maximum vertical velocity: decay indicates a stable flow, and exponential growth—unstable flow. Plot the growth rate versus Ra, and use your favorite scripting language to fit (e.g. using least squares) a line to these points. The root of the fitted function is the estimate for $Ra_{crit.}$.

References

1. Barakat, H.Z., Clark, J.A.: On the solution of the diffusion equations by numerical methods. J. Heat Transf. **88**(4), 421–427 (1966)
2. Bouzidi, M., d'Humieres, D., Lallemand, P., Luo, L.S.: Lattice Boltzmann equation on a two-dimensional rectangular grid. J. Comput. Phys. **172**(2), 704–717 (2001)
3. Buckingham, E.: On physically similar systems; illustrations of the use of dimensional equations. Phys. Rev. **4**(4), 345–376 (1914)
4. Budyko, M.I.: The effect of solar radiation variations on the climate of the Earth. Tellus **21A**(5), 611–619 (1969)
5. Chen, D., Gerdes, R., Lohmann, G.: A 1-D atmospheric energy balance model developed for ocean modelling. Theor. Appl. Climatol. **51**(1–2), 25–38 (1995)
6. Chen, S., Doolen, G.D.: Lattice Boltzmann method for fluid flows. Annu. Rev. Fluid Mech. **30**(1), 329–364 (1998)
7. Chopard, B., Droz, M.: Cellular Automata Modelling of Physical Systems. Cambridge University Press, Cambridge (1998)
8. Courant, R., Friedrichs, K., Lewy, H.: Über die partiellen Differenzengleichungen der mathematischen Physik. Math. Ann. **100**(1), 32–74 (1928)
9. Courant, R., Friedrichs, K., Lewy, H.: On the partial difference equations of mathematical physics. IBM J. Res. Dev. **11**(2), 215–234 (1967)
10. Dellar, P.J.: Incompressible limits of lattice Boltzmann equations using multiple relaxation times. J. Comput. Phys. **190**(2), 351–370 (2003)
11. Dellar, P.J.: An interpretation and derivation of the lattice Boltzmann method using Strang splitting. Comput. Math. Appl. **65**(2), 129–141 (2013)
12. Ginzburg, I., d'Humieres, D.: Multireflection boundary conditions for lattice Boltzmann models. Phys. Rev. E **68**(6), 066614 (2003)
13. Guo, Z., Shu, C.: Lattice Boltzmann Method and Its Applications in Engineering. World Scientific Publishing Co., Singapore (2013)

14. Guo, Z., Zheng, C., Shi, B.: Discrete lattice effects on the forcing term in the lattice Boltzmann method. Phys. Rev. E **65**(4), 046308 (2002)
15. Haney, R.L.: Surface thermal boundary condition for ocean circulation models. J. Phys. Oceanogr. **1**(4), 241–248 (1971)
16. Kuo, L., Chen, P.: Numerical implementation of thermal boundary conditions in the lattice Boltzmann method. Int. J. Heat Mass Transf. **52**(1–2), 529–532 (2009)
17. Lohmann, G., Gerdes, R., Chen, D.: Stability of the thermohaline circulation in a simple coupled model. Tellus **48A**(3), 465–476 (1996)
18. Mohamad, A.A.: Lattice Boltzmann Method: Fundamentals and Engineering Applications with Computer Codes. Springer, London (2011)
19. Nakamura, M., Stone, P.H., Marotzke, J.: Destabilization of the thermohaline circulation by atmospheric eddy transports. J. Clim. **7**(12), 1870–1882 (1994)
20. Pletcher, R.H., Tannehill, J.C., Anderson, D.: Computational Fluid Mechanics and Heat Transfer. CRC Press, Boca Raton (2012)
21. Prange, M., Lohmann, G., Gerdes, R.: Sensitivity of the thermohaline circulation for different climates—investigations with a simple atmosphere-ocean model. Paleoclimates **2**(1), 71–99 (1997)
22. Press, W.H., Teukolsky, S.A., Vetterlin, W.T., Flannery, B.P.: Numerical Recipes in Fortran 77, 2nd Edn. Volume 1: The Art of Scientific Computing. Cambridge University Press (1992). also available as http://apps.nrbook.com/fortran/index.html
23. Rahmstorf, S.: On the freshwater forcing and transport of the Atlantic thermohaline circulation. Clim. Dyn. **12**(12), 799–811 (1996)
24. Robinson, J.A.: Software Design for Engineers and Scientists. Elsevier, United Kingdom (2004)
25. Rooth, C.: Hydrology and ocean circulation. Prog. Oceanogr. **2**(11), 131–149 (1982)
26. Stommel, H.: Thermohaline convection with two stable regimes of flow. Tellus **13A**(2), 224–230 (1961)
27. Strang, G.: Computational Science and Engineering. Wellesley-Cambridge Press, Wellesley (2007)
28. Succi, S.: The Lattice Boltzmann Equation: for Fluid Dynamics and Beyond. Oxford University Press, Oxford [u.a.] (2001)
29. Sukop, M.C., Thorne, D.T.: Lattice Boltzmann Modeling: An Introduction for Geoscientists and Engineers. Springer, Berlin [u.a.] (2006)
30. Tritton, D.J.: Physical Fluid Dynamics. Clarendon Press; Oxford University Press, Oxford [England]; New York (1988)
31. Wang, J., Wang, D., Lallemand, P., Luo, L.S.: Lattice Boltzmann simulations of thermal convective flows in two dimensions. Comput. Math. Appl. **65**(2), 262–286 (2013)
32. Wolf-Gladrow, D.A.: Lattice-gas Cellular Automata and Lattice Boltzmann Models: An Introduction. Springer, New York (2000)
33. Yu, D., Mei, R., Luo, L.S., Shyy, W.: Viscous flow computations with the method of lattice Boltzmann equation. Prog. Aerosp. Sci. **39**(5), 329–367 (2003)

Chapter 5
More Advanced Techniques

In this chapter, we introduce several techniques and tools (build systems, more efficient I/O, parallelization, etc.), which are commonly used in ESS applications. Most of these concepts are also relevant for other programming languages. With such a large list of topics, it is clearly impractical to be comprehensive. Nonetheless, through the examples, we hope to provide the reader with a reasonable overview of how these facilities can be used, and some intuition about how they can be combined.

5.1 Multiple Source Files and Software Build Systems

Each of the examples provided so far consisted of a single source file, which contained the code for the main-program and for any accompanying modules and procedures. To obtain the final executable, we simply compiled the file manually (see Sect. 1.3). While this approach is often acceptable for small test programs, it becomes inconvenient for large applications. A separation of the code into several files (potentially arranged into a multi-level directory hierarchy) is preferred instead, for a variety of reasons:

- a single file would become too large to comprehend—multiple files can improve readability when they are used to demarcate sub-components of the application (especially when using OOP)
- code reuse (within the application and across multiple application), as well as collaboration in teams are greatly simplified
- in combination with a software build system (and with some planning), this approach can prevent compilation times from increasing too much

A price to be paid for these benefits, however, is a more complex compilation process: whereas in the previous examples we could let the compiler handle transparently the compilation and linking stages, with the multiple source file approach the programmer needs to be aware of the intermediate object files, libraries, etc. We briefly review these topics in the next section.

© Springer-Verlag Berlin Heidelberg 2015
D.B. Chirila and G. Lohmann, *Introduction to Modern Fortran for the Earth System Sciences*, DOI 10.1007/978-3-642-37009-0_5

DISCLAIMER

Unfortunately, the procedures for creating/using object files and libraries are not standardized—this is what makes portable software development with compiled languages difficult. We recommend to check the documentation of your OS and of your compiler, and adjust the steps in this section accordingly. For brevity, we focus mostly on the Linux system, with the gfortran compiler.

5.1.1 Object Files, Static and Shared Libraries

5.1.1.1 Object Files

We already mentioned object files in Sect. 1.3. Each of these files (with extension `.o` in Linux and OSX and `.obj` in windows) contains the machine code generated from the corresponding source code file (after *compilation* and *assembly*), but without any code from other libraries. With the GNU compilers, these files are created by passing the `-c` option at the compilation command line. For example, assuming we have three source code files named `util1.f90`, `util2.f90` and `main.f90` (where the first two contain modules and/or procedures which are used in the third one, where the main-program resides), we can produce the corresponding object files with:

```
$ gfortran -c util1.f90    # produces object file util1.o
$ gfortran -c util2.f90    #                      util2.o
$ gfortran -c main.f90     #                      main.o
```

Assuming that these are the only object files in our application, we can *link* them into an executable. This step is also initiated by the compiler (when invoked with the `-o <executable_name>` option[1]):

```
$ gfortran -o main main.o util1.o util2.o
```

Working with more than one compiler

A peculiarity of Fortran is that, for programs which use modules, many compilers will also produce intermediate files with the `.mod` extension (one for each module). The role of these files is similar to that of header files in C

---

[1] This flag can actually be omitted, in which case the executable name would default to `a.out` (not too informative).

and C++ (i.e. helping the compiler to check interfaces). However, a significant difference is that .mod files are generated automatically, and usually not portable between different compilers (or even between different releases of the same compiler). Therefore, it is best to avoid mixing code obtained with different compilers. This implies that, when switching compilers, we have to re-compile not only our program, but also the libraries which our code uses. It is often necessary in such cases to tell each compiler where it can find the .mod files for its corresponding version of a compiled library. Many compilers allow this with the -I<path_to_mod_files> directive; this needs to be used in addition to the -L compiler option, which will be discussed below.

5.1.1.2 Static Libraries

In theory, it is possible to re-use third-party software by directly listing multiple object files at the command line when linking takes place. However, this would lead to a proliferation of object files in the file system and very long lists of files for the linking phase, which is inconvenient in practice. It is better, instead, to package the object files together. Static libraries (usually with extension .a in Linux and .lib in windows) were the first approach developed for achieving this. Following the example above, we could combine util1.o and util2.o into a static library. In Linux, this is done with the ar command, as in:

```
$ ar rscv libutil.a util1.o util2.o
```

Check the man-page of ar for more information about the command line options, and about the other operations which are possible.

There are two equivalent methods for making the newly created static library available to the linker when our final executable is created. The first method (useful mostly when the library is used only internally within a project[2]) consists of simply adding the library name to the list of files to be passed to the linker:

```
$ gfortran -o main main.o libutil.a
```

The second method (handy for libraries needed by many applications) consists of two sub-steps:

1. If necessary, the directory where the library resides is added (using the -L<path_to_dir> option) to the list of directories where the linker will search for libraries. This step may not be necessary if the static library was

---

[2] The term *convenience library* is also commonly used to denote such a scenario.

installed in standard path. For our current example, we would add the current
directory, so the option would become $\boxed{\texttt{-L\$PWD}}$.

2. Also as an option to the linker, the name of the library is added, by using
the $\boxed{\texttt{-l<abbrev\_lib\_name>}}$. Here, `<abbrev_lib_name>` stands for the
name of the library file, from which the $\boxed{\texttt{lib}}$ in the front and the extension
($\boxed{\texttt{.a}}$ for static libraries) are dropped. In our example, based on the library name
$\boxed{\texttt{libutil.a}}$ the option to be passed to the linker becomes $\boxed{\texttt{-lutil}}$.

Combined, the second method for presenting libraries to the linker becomes for our
three-files project:

```
$ gfortran -o main main.o -L$PWD -lutil
```

This second approach is recommended when using libraries installed in system fold-
ers, where the linker would search by default. Also, if a shared library with that name
is found by the linker, it will use that,[3] for efficiency reasons.

When an object file or a static library (using either of the methods above) is made
available to the linker, it will select the entities (procedures, data, etc.) which the
application refers to and just copy them inside the executable. The executable and
the object/library files can then go their separate ways—for example, we could delete
the libraries and we would still be able to execute the program. This decreases the
dependencies on external packages, which is particularly useful when the application
is distributed/deployed to users in binary form.

However, there are also some scenarios where this type of libraries are not well-
suited, due to some serious disadvantages. Let us assume that you have written a
very useful library, and that many developers want to use it in their own programs.
However, making the library static brings along some disadvantages:

• From an efficiency point of view, static libraries are plagued by duplication of code,
which shows up in various places. Depending on the size of the library and on the
number of programs using it, any of the following issues can become significant:

  – If users commonly install several programs which use your library, the code
  from your library will be duplicated several times on disk.
  – In addition, if several programs which use your library are running at the same
  time, the same duplication will appear in memory, when the program is executed.
  Besides wasting resources again, this can cause various kinds of performance
  problems (e.g. applications taking a long time to start, or performing poorly, due
  to instruction cache misses). However, to be fair, there are also some situations
  where use of static libraries can outperform shared ones (especially when entities
  from the library are accessed in time-consuming loops).

---

[3] Many compilers still offer the option to override this behavior, if the developer insists on static
linking; in `gfortran`, the $\boxed{\texttt{-static}}$ flag can be used for this (or $\boxed{\texttt{-Wl,-Bstatic}}$ and
$\boxed{\texttt{-Wl,-Bdynamic}}$ to toggle static linking on and off for specific libraries).

- If you develop an updated version of the library (perhaps to fix a bug or to improve performance), all the programs using your library have to be re-compiled and re-distributed to users. This usually makes updates very slow to propagate throughout the userbase.

5.1.1.3 Shared Libraries

To overcome most of the disadvantages discussed above, shared libraries were developed. Depending on your OS, you may also encounter these named as *shared objects*, *dynamically linked libraries*, *frameworks*, *dynamically linkable libraries*, etc. The details of how these are created and used are (unfortunately) highly OS-specific. However, the basic idea is the same: instead of copying the library code directly in the executables, only the names of the libraries that will be needed by the executable are recorded inside. When a user eventually tries to run the executable, a system component known as the *dynamic linker* will match the libraries needed by the executable against what is available on the system. This happens before any of the program's code is actually executed. If the dynamic linker cannot satisfy all the requirements of the executable, it will usually cause the entire program to abort, with an error message.

Shared libraries can solve most of the problems that plagued static libraries, precisely because they use this extra level of indirection at runtime. Only one copy of the library is required on disk, no matter how many executables need this code. Also, assuming we want to run several programs that all use a certain library, the library code will need to be loaded in memory only for the first program—the OS will then make this code available[4] to the other programs which need it, saving both space in memory and time (since no re-loading is necessary).[5]

Creating shared libraries To re-use our three-files example from above, we could create a shared library (using the GNU compiler) from the files util1.f90 and util2.f90 in two steps:

1. When creating the initial object files, the -fPIC flag is required by gfortran, as in:

```
$ gfortran -fPIC -c util1.f90      # produces object file util1.o
$ gfortran -fPIC -c util2.f90      #                      util2.o
```

The additional compiler flag enables generation of object code which is said to be *position-independent*, which is necessary for enabling true sharing of the library code (see, e.g., Calcote [3] for details).

---

[4] Only the code is shared—data entities declared by the library are private to each of the programs.

[5] We constructed here a positive picture of shared libraries. In practice, things can be "spoiled" by the potential existence of different versions of the same library (see e.g. Hook [10]). For badly-designed libraries, these problems can outweight the possible benefits.

2. The second step is to create the shared library itself. In contrast to the static libraries, this step is usually performed through the compiler. For our example, we can use:

```
$ gfortran -shared -o libutil.so util1.o util2.o
```

There are many other subtleties related to designing, creating and maintaining shared libraries, which exceed the scope of our basic introduction (for the interested reader, we suggest Hook [10] as a general discussion, or Kerrisk [11] for Unix-specific information). Instead, we focus below on how to use shared libraries written by someone else, since this topic is a common source of frustration (especially for beginners). This discussion is also relevant for Sect. 5.2.2, which demonstrates how to work with the netCDF-library (which implements operations on a very popular data storage format in ESS).

Using shared libraries Two contexts need to be considered when working with a shared library—the *link-time* and *runtime*.

The first phase (at *link-time*) causes the names of the shared libraries needed by our application to be recorded within the executable. The syntax for performing this step is actually almost the same as for static libraries—the only difference is that, if we choose to specify the full path, we generally need to use different file extensions—usually $\boxed{.so}$[6] in Linux, $\boxed{.dylib}$ in OSX or $\boxed{.dll}$ in Windows. If the library is not in a system-wide path, we can add the appropriate directory to the list of paths inspected by the linker. For our example, this would lead to the following command to produce the final executable:

```
$ gfortran -o main main.o -L$PWD -lutil
```

The second phase (at *runtime*) starts when a user issues the command to execute our application. Even before any code from the application is run, the *dynamic linker* will locate all shared libraries needed by our application and let our application know how it can access them. This step can often fail, especially if the program depends on shared libraries which are not installed system-wide (in the places where the dynamic linker usually searches). For example, the executable we produced above may fail to execute with the error:

```
$ ./main
./main: error while loading shared libraries: libutil.so: cannot open shared
        object file: No such file or directory
```

The dynamic linker does not know where to find the "util" library, and causes the whole program to abort.

Our readers using Linux may have encountered similar errors with other applications, although the error can occur on any OS.

---

[6] This is an abbreviation which stands for "shared object".

Some very useful tools exist for checking which shared libraries are needed by an executable, and if these would be found by the dynamic linker. On `Linux`, the tool to use is `ldd`. In our case, this would report something like:

```
$ ldd ./main
...
libutil.so => not found
...
```

Equivalent functionality is available in other systems (e.g. `otool -L` in OSX, the `Dependency Walker`[7] program in `Windows`), etc.

Of course, once we identify such a runtime linking problem, the question is how to fix it. The following options are available when working on a `Linux` system:

1. If the library is installed system-wide, this is a problem to be solved by the system administrator (by checking that the appropriate package is installed and/or adding the path to the library's directory, using the `ldconfig` program).

2. It is also possible to solve the problem from the user's side. This is recommended, for example, when we are dealing with an experimental library, not relevant to other users. Several approaches are possible:

   - It is possible to encode the non-standard library path directly into the executable, by passing some additional options to the linker. In our example above, we could accomplish this with the command[8]:

   ```
   $ gfortran -o main main.o -Wl,--enable-new-dtags -Wl,-rpath,$PWD -L$PWD -lutil
   ```

   Here, the `-Wl,-rpath,$PWD` part will cause the linker to add the working directory[9] to the list of libraries hard-coded within the executable. As long as these required libraries are not removed, our program will run without any further interventions.

   The `-Wl,--enable-new-dtags` relates to issues of priority of library paths. Without this option, the default (but deprecated) effect on `Linux` will be to give higher priority to the paths within the executable. However, the recommended approach nowadays (with this option specified) gives higher priority to paths in the `LD_LIBRARY_PATH` environment variable (discussed next). For example, this allows developers to test a program with a new version of a library (without re-compiling the program).

---

[7] http://www.dependencywalker.com.

[8] Note that in general you will have to replace `$PWD` to reflect the path of your library.

[9] Again, other paths can be specified insted—for example, assuming we have some custom libraries in `/home/my_username/libs`, we could use `-Wl,-rpath,/home/my_username/libs`.

- As an alternative, many authors recommend to add the directory containing the library to the user's LD_LIBRARY_PATH variables (DYLD_ LIBRARY_PATH on OSX, LIBPATH on IBM's AIX, or the PATH variable on Windows[10]). This approach works, but we recommend to use it on a per-program basis (shell aliases can be used to make the invocations shorter), to avoid polluting the user's environment.[11] With these pitfalls in mind, we could use this approach to allow our test program to run:

```
$ env LD_LIBRARY_PATH=$LD_LIBRARY_PATH:$PWD ./main
```

As a final remark on shared libraries, it is worth mentioning that most systems offer an even more advanced (but also more flexible) mechanism known as *dynamic linkage*, whereby applications can have full control on the library-loading mechanism (i.e. they can search for arbitrary libraries at runtime and execute code from them). Many applications use this powerful facility to add support for plugins—for example web browsers.

5.1.2 Introduction to GNU Make (gmake)

From the previous section, it may be clear to the reader that the build process (including creation and use of libraries) can become quite complex for nontrivial projects. Although there is sometimes educational value in walking through the steps for building a project (especially when debugging build problems), it would certainly be a bad use of human resources to type all commands every time a source file is modified—computers are much better suited for these tasks. Therefore, many techniques and tools were developed to automate this process, as well as other repetitive tasks which occur in the software development workflow (running of automatic tests, preparation of final user-installable packages, etc.). In terms of complexity and built-in functionality, these tools range from simple scripts (see Sect. 5.6.1) to advanced build systems such as autoconf+automake+libtool,[12] CMake, or SCons.

In this section we focus on *GNU Make* (gmake),[13] which is an intermediate-complexity build system that is sufficient in many cases. Although readers may eventually use a different build system, some basic familiarity with make is instructive, since this system encourages thinking explicitly through the basic actions nec-

---

[10] In Windows, the linker will usually also search in the directory where the executable program resides.

[11] Adding paths to this variable in the user's shell configuration files can cause performance and security problems, in addition to hurting portability.

[12] These are commonly used together, so they are referred to as autotools.

[13] For the sake of brevity, we use the make acronym sometimes, but always imply gmake, which is easily available on Unix-systems (other systems may refer to this as gmake).

essary for creating the final products (whereas other systems may hide some of these details).

To a first approximation, we can think of make as a program which automatically constructs some output files from a set of input files, based on some recipes. The input files are, in most programming projects, the actual source code created by the programmer, while the output files are often the compiled executable programs. Using the jargon of make, we call the former (input) *dependencies* or *prerequisites*, and the latter (output) *targets*. However, to extend this picture, both *targets* and *dependencies* can also be tasks in a more abstract sense, because not all work that is automated with make fits the file-transformation model.

Since each project is generally unique, it is the task of the programmer to describe to make how the various entities (source code files, object files, libraries, data files, executable, etc.) depend on each other. These dependencies are known as *rules* by make. Since any entity can take the role of target in one rule and of prerequisite in another rule,[14] it is useful to think of rules as links in one or more *directed acyclic graphs* DAGs of dependencies. It is the task of make to construct internal representations of such graphs and, afterwards, to traverse the links as appropriate for correctly updating the current target. Of course, to actually perform tangible actions, there are usually (but not always) some *commands* which are associated to specific rules. Here, we should point out that make also has an internal database of rules, many of which are expressed as generic patterns, so a command may become associated to a specific rule, even if no command was explicitly specified by the developer.

To assimilate rules and target inter-dependencies specific to the project, make searches in the current working directory for a file named GNUmakefile, makefile, or Makefile (in that order). For brevity, we will refer to these collectively as just makefiles. Many projects have a single such file, and this is also the scenario we will assume in our examples here. However, the situation can quickly become very complex, especially if the code is spread across several directories (see Mecklenburg [20] for a make-specific discussion, or Smith [26] for some useful perspective on these matters).

5.1.2.1 Invoking **Make**

Assuming there is a makefile in the present working directory, the build process can be started by simply typing at the command line:

```
$ make
```

or, if more control over the targets to build is required, by passing as an argument a space-separated list of targets, as in:

---

[14] Intermediate object files are a fitting example, since they are created when source code is compiled, and consumed when the executables are eventually prepared.

```
$ make target1 target2
```

There are many command line options for customizing the behaviour of make in useful ways—for example:

- If, for some reason, the project uses a non-standard name for the makefile, which make does not recognize by default, we can point it to the custom file, using the $-f$ flag. This can happen, for example, when the project needs to be compiled with multiple toolchains (compilers, linkers, etc.)—a possible make-based solution to this problem would be to provide a different makefile for each platform, and let the user select a specific one, as in:

```
$ make -f Makefile_MachineX_ToolchainY
```

- Another time-saving feature of make is "parallel build", which distributes sub-branches of the dependency graph, for concurrent execution on the multiple cores of modern CPUs. The flag which enables this option is $-j$, followed by a number (of execution threads), as in:

```
$ make -jX
```

where x should be close to the count of CPU-cores in the machine (a common rule of thumb is to use $x = n_{cores} + 1$, so use $-j5$ for a machine with four processing cores).
- It is sometimes necessary to debug errors in makefiles themselves. The $-n$ flag, which causes make to only display the commands it would otherwise execute, is useful in such cases.
- As a final invocation option that we mention here, the $-p$ flag on its own causes make to simply print its internal database of rules and environment variables, which can be useful for debugging makefiles and for identifying common patterns which could be used to our advantage.

5.1.2.2   How Make Processes Files

As mentioned already, when make is invoked it usually proceeds by performing two passes through the file system (see, e.g., Calcote [3] for more details).

First, from the makefile(s) of the project and any matching rules from its internal database, make synthesizes corresponding dependency graphs and saves these into its internal runtime data structures. Also at this stage, variables are resolved.

For the second pass, make deals with the specific *target(s)* that need to be built. The selection of top-level targets can be dictated by the user at the command line (when invoking make). If no preference was expressed by the user, the default behaviour is to build the first *target* encountered while scanning the makefile(s) in the previous phase. After the top-level targets have been selected, `make` processes each one, by taking it as the root node of a dependency graph traversal, descending (recursively) to the leaf nodes, and then making its way back to that root node and executing the appropriate *commands* whenever some *target* is found to be older than any of its *dependencies* (as ascertained based on the modification times tracked by the underlying file system).

For example, if we have an executable `my_program` which depends on the file `my_program.f90`, make compares the modification times of the two files, and re-creates the former whenever it finds that it either (a) is older than the source file or (b) is missing. This criterion causes `make` to perform the minimal amount of work necessary to update a target, avoiding a lot of extraneous re-compilation when only a few files have changed (which is the typical scenario).

5.1.2.3 Example: Using **Make** with the Climate Box Model Application

In Sect. 4.2, we discussed the implementation of an inter-hemispheric climate box model. The code itself was all placed in a single file (see file `src/Chapter4/box_model_euler.f90`, in the repository). Here, we will demonstrate some additional features of make, by distributing the various components of that example across several files. The process of distributing the code is actually straightforward in this case, since the code was already organized into several modules (these are `NumericKinds`, `PhysicsConstants`, `ModelCons-tants`, `GeomUtils` and `ModelState_class`). We simply create a new file for each module,[15] and rename the file containing the main-program, to `box_model_euler_main.f90`, to emphasize its role. No code changes are necessary. Of course, the more interesting part as far as this section is concerned is the makefile. In fact, we will write several versions of the makefile, starting from a very simple (but also unreasonably verbose) one. Later versions then leverage make features for improving the build specification.

Explicit rules and basic `makefile`**-layout** Without further ado, our initial version of the makefile (which would cause make experts to frown) is:

```
 5  box_model_euler: box_model_euler_main.o ModelState_class.o \
 6      ModelConstants.o PhysicsConstants.o \
 7      GeomUtils.o NumericKinds.o
 8      gfortran -o box_model_euler \
 9          box_model_euler_main.o ModelState_class.o \
10          ModelConstants.o PhysicsConstants.o \
11          GeomUtils.o NumericKinds.o
12
13  box_model_euler_main.o: box_model_euler_main.f90
14      gfortran -c box_model_euler_main.f90
15
16  ModelState_class.o: ModelState_class.f90
17      gfortran -c ModelState_class.f90
```

---

[15] This is simply a convention, similar to the general recommendation of having one class for each `.h—.cpp` pair in C++.

```
18
19  ModelConstants.o: ModelConstants.f90
20      gfortran -c ModelConstants.f90
21
22  PhysicsConstants.o: PhysicsConstants.f90
23      gfortran -c PhysicsConstants.f90
24
25  GeomUtils.o: GeomUtils.f90
26      gfortran -c GeomUtils.f90
27
28  NumericKinds.o: NumericKinds.f90
29      gfortran -c NumericKinds.f90
30
31  # additional dependencies
32  box_model_euler_main.o ModelState_class.o:NumericKinds.o
33  ModelConstants.o PhysicsConstants.o GeomUtils.o: NumericKinds.o
34
35  box_model_euler_main.o: PhysicsConstants.o ModelConstants.o
36  box_model_euler_main.o: ModelState_class.o
37
38  ModelConstants.o GeomUtils.o: PhysicsConstants.o
39  ModelState_class.o: GeomUtils.o
```

Listing 5.1 `src/Chapter5/BoxModelMultipleFiles/Makefile.v1`
(excerpt)

Let us discuss the various elements in this file. To get the less interesting syntax features out of our way first, note that all text after a hash mark (# , as in *line 32*) is treated as a comment, akin to most Unix shells.[16] Another feature in common with Unix shells is the way text can "spill" over multiple lines, by appending a backslash (\ ) at the end of each line to be continued (*lines 6–12* above). This is roughly equivalent to the & -character used for extending a line of code in Fortran.

Although it looks dense, the structure of the file is quite simple: it consists entirely of what are known as *explicit rules* (at *lines 5–11, 13–14, 16–17, 19–20, 22–23, 25–26, 28–29, 32, 33, 35–36, 38* and *39*). make does not require every rule to have commands associated with it and, indeed, the last five rules above (*lines 32–39*) are like this.

For a rule which *does* have associated commands (like the second rule, at *lines 13–14*), note that the command lines need to be indented with an *explicit* TAB character (NOT spaces), to demarcate the commands.[17] If this rule is not followed, make will probably fail to build, often with cryptic errors.

The rules themselves are not surprising: to compile the final executable (first rule, at *lines 5–11*), we list all the object files as prerequisites, followed by the command which invokes the linker (*lines 8–11*). The next six rules, which specify how to create each object file, are very similar to each other—only the *stem* (i.e. filename without extension) of the filename is changing. Finally, *lines 32–39* specify some additional dependencies, which are mostly dictated in our case by the way Fortran modules include each other in the various files of our project. In particular, *lines 32–33* specify that all other object files depend on NumericKinds.o , since most of the code depends (directly or indirectly) on this module.

---

[16] An exception to this rule is when the hash is embedded within a command—commands are passed as they are to the shell (including hashes).

[17] This may require additional configuration for some text editors.

Finally, note that *lines 35–36* have the same target (box_model_euler_ main.o). When make scans our makefile, it will actually combine these two lines into a single rule. It is often useful (and clearer) to specify a rule with many prerequisites in several pieces, like this. However, only one of these "sub-rules" can have commands attached to it, since there should not be more than one way to make the same target.

Pattern rules, wildcards, and automatic variables As already mentioned, there is much room for improvement in our previous makefile. *Lines 13–29* are a good "offender" to tackle first. There, the reader may recognize that the same pattern is repeated six times (only the filename changes). The next version of our makefile generalizes these rules:

```
 5  box_model_euler: box_model_euler_main.o ModelState_class.o \
 6       ModelConstants.o PhysicsConstants.o \
 7       GeomUtils.o NumericKinds.o
 8       gfortran -o box_model_euler \
 9           box_model_euler_main.o ModelState_class.o \
10           ModelConstants.o PhysicsConstants.o \
11           GeomUtils.o NumericKinds.o
12
13  %.o: %.f90
14       gfortran -c $<
15
16  # additional dependencies
17  box_model_euler_main.o ModelState_class.o: NumericKinds.o
18  ModelConstants.o PhysicsConstants.o GeomUtils.o: NumericKinds.o
19
20  box_model_euler_main.o: PhysicsConstants.o ModelConstants.o
21  box_model_euler_main.o: ModelState_class.o
22
23  ModelConstants.o GeomUtils.o: PhysicsConstants.o
24  ModelState_class.o: GeomUtils.o
```

Listing 5.2 | src/Chapter5/BoxModelMultipleFiles/Makefile.v2 |
(excerpt)

The new code (*lines 13–14* in Listing 5.2) replaces the explicit rules in *lines 13–29* of Listing 5.1. In fact, if we would later add Fortran files to our project, the new rule would be able to build the corresponding object file automatically[18] (with the previous approach, we would need to remember to add yet another explicit rule).

To understand the new code, note that a percent character (%) acts as a place-holder, which matches any number of any characters. When make encounters such a pattern rule, it will remember it, and try to use it whenever it encounters a target that it would not know how to build otherwise. Here, our pattern rule "teaches" make how to produce any object file (with the .o extension) from the corresponding source code file (with the same stem, but with the .f90 extension), if the latter exists.

Let us analyze now the actual command (*line 14*), which is executed whenever the pattern rule matches. From our discussion in Sect. 5.1.1, the gfortran −c part should look familiar: the compiler is invoked, with the flag for only compiling code, without linking in anything from external libraries. But which file are we compiling?

---

[18] In case there are exceptions that should not be built like this, make also supports *static pattern rules* (see Mecklenburg [20])—these are basically pattern rules, with scope restricted to a certain (user-controllable) list of files.

This is specified by the $\boxed{\$<}$ part, which is an example of what make calls *automatic variables*. These variables are automatically assigned internally by make, whenever a match of a rule is found, and their scope is restricted to the commands associated with the rule (if any). make stores into these automatic variables information about the specific target and prerequisite(s) that the rule matched. This is crucial for writing generic commands, to be associated with pattern rules.

The specific automatic variable that we used above ($\boxed{\$<}$) is expanded in the command to the *filename of the first prerequisite* which is, in our case, the Fortran source code file we wanted to compile. Other interesting automatic variables (see Mecklenburg [20] or make's documentation for a comprehensive list) are:

- $\boxed{\$@}$ —*name of the current target*
- $\boxed{\$\char`^}$ —*space-separated list of all prerequisites, with duplicates removed*
- $\boxed{\$+}$ —*same as above, but keeping the duplicates*

The dollar sign is actually not part of the name of automatic variables—it is an operator which expands ("dereferences") the value of the variable. This syntax also holds for normal variables, which we will demonstrate soon.

Normal variables Looking at the previous code listing, we notice that there is still some duplication for the first rule (the names of the object files are written twice in *lines 5–11* of Listing 5.2). We already advocated for reducing code duplication (as one of the ways to make software more robust), so let us do that here too. As before, we provide the code first, and explain it later:

```
6   srcs := box_model_euler_main.f90 ModelState_class.f90 \
7        ModelConstants.f90 PhysicsConstants.f90 \
8        GeomUtils.f90 NumericKinds.f90
9   objs := $(srcs:.f90=.o)
10  prog := box_model_euler
11
12  $(prog): $(objs)
13       gfortran -o $@ $^
14
15  %.o: %.f90
16       gfortran -c $<
17
18  # additional dependencies
19  box_model_euler_main.o ModelState_class.o: NumericKinds.o
20  ModelConstants.o PhysicsConstants.o GeomUtils.o: NumericKinds.o
21
22  box_model_euler_main.o: PhysicsConstants.o ModelConstants.o
23  box_model_euler_main.o: ModelState_class.o
24
25  ModelConstants.o GeomUtils.o: PhysicsConstants.o
26  ModelState_class.o: GeomUtils.o
```

Listing 5.3 $\boxed{\text{src/Chapter5/BoxModelMultipleFiles/Makefile.v3}}$
(excerpt)

First, in *lines 6–8*, we instruct make to store a list of all source code files of our project, into the variable $\boxed{\text{srcs}}$. Note that, unlike Fortran or other languages, make does not require us to specify the type of variables; indeed, that would be pointless, since there really is only one type in make, namely character strings (usually containing filenames separated by spaces[19]). When a variable appears in the LHS of an

---

[19] Unfortunately, this can drastically complicate things on Windows, if the file paths/names contain spaces. Therefore, if using make on Windows, we recommend to avoid such "non-Unix-friendly"

assignment operator, the name of the variable is written normally. However, when we want to use (expand) the value of the variable in another place, we usually need to surround the variable's name by braces, and precede the resulting construct by a dollar sign (see, e.g., *line 12 above*).[20]

As a slight twist, we did not remove the duplication in the previous `makefile` (Listing 5.2) by saving the list of object files into a variable. Instead, we saved in `srcs` the list of `.f90` files which, being the leafs of the entire dependency graph, provide a more natural starting point. Then, in *line 9* of Listing 5.3, we use a handy make feature, which "maps" the list of source files onto the list of corresponding object files (`objs`). Finally, in *line 10* we introduce `prog`, which holds the name of our final executable.

With these new variables (and using more of the automatic variables), the rule for linking the executable becomes much more readable (compare *lines 12–13* above with *lines 5–11* in Listing 5.2).

Before proceeding with other topics, a few words are in order, regarding the assignment operators in `make`. The type of assignment we used above (with the `:=` operator) leads to an *immediate* evaluation of the expression on the RHS. `make` also supports *recursively expanded variables* (created with the `=` operator), which are evaluated only when `make` actually needs the value for proceeding with its work. We leave this as a topic of further exploration for the interested reader (see e.g. Mecklenburg [20]).

Improving portability, overriding values at the command line, and phony targets
As a final iteration on our example, we can change a few things in the `makefile`, to make it more easily portable to other systems. For example, in Listing 5.3, we hard-coded (at *lines 13* and *16*) the commands for compiling and linking the components of our program. If we needed to use another compiler or different compiler options, we would need to change the `makefile` accordingly. However, `make` provides a set of intrinsic variables and rules, which we can leverage to make our `makefile` more user-friendly, as demonstrated below:

```
22   srcs := box_model_euler_main.f90 ModelState_class.f90 \
23          ModelConstants.f90 PhysicsConstants.f90 \
24          GeomUtils.f90 NumericKinds.f90
25   objs := $(srcs:.f90=.o)
26   prog := box_model_euler
27
28   $(prog): $(objs)
29       $(LINK.f) $^ $(LOADLIBES) $(LDLIBS) $(OUTPUT_OPTION)
30
31   %.o: %.f90
32       $(COMPILE.f) $< $(OUTPUT_OPTION)
33
34   clean:
35       -$(RM) *.mod *.o $(prog)
36
37   .PHONY: clean
38
```

(Footnote 19 continued)
paths. If such a compromise is not acceptable, switching to another build system such as *Cross Platform Make* (CMake) or the *Software Construction tool* (SCons) may be a more fruitful strategy.

[20] Exceptions are the automatic variables (discussed previously), where the brackets can be (and usually are) omitted, since they consist of a single letter.

```
39   # additional dependencies
40   $(filter-out NumericKinds.o, $(objs)):NumericKinds.o
41
42   box_model_euler_main.o: PhysicsConstants.o ModelConstants.o
43   box_model_euler_main.o: ModelState_class.o
44
45   ModelConstants.o GeomUtils.o: PhysicsConstants.o
46   ModelState_class.o: GeomUtils.o
47
48   # WARNING: next two values are specific to the GNU compiler -- readers should
49   #          adjust this if they are using another compiler/compiler-version.
50   FC := gfortran-4.8
51   FFLAGS := -O2 -std=f2008ts -pedantic -Wall
```

Listing 5.4 | `src/Chapter5/BoxModelMultipleFiles/Makefile.v4` |
(excerpt)

The most important changes compared to the previous makefile appear in *lines
29* and *32* of the new Listing 5.4. Here, we use the intrinsic variables $\boxed{\text{LINK.f}}$,
$\boxed{\text{COMPILE.f}}$, $\boxed{\text{LOADLIBES}}$, $\boxed{\text{LDLIBS}}$, and $\boxed{\text{OUTPUT\_OPTION}}$ instead of
directly hard-coding the program names. These variables are examples of the *recur-
sively expanded variables* we already mentioned. Here, the nature of the variables
matters, because they allow us to provide a specific compiler at any point in the
makefile. Specifically, at *lines 50* and *51*, we define variables $\boxed{\text{FC}}$ and $\boxed{\text{FFLAGS}}$,
which usually stand for *Fortran compiler* and *Fortran compiler flags*, respectively.
make will use these to construct $\boxed{\text{COMPILE.f}}$ and $\boxed{\text{LINK.f}}$, when the time will
come to evaluate those variables.

A very convenient feature of make variables is that we can override them directly
from the command line,[21] when we invoke make. This is achieved by providing
a list of variable assignments; for example, the following invocation would cause
gfortran to use more aggresive optimizations (and to disable warnings), in contrast to what we specified in the makefile:

```
$ make -f Makefile.v4 FFLAGS='-O3'
```

This feature is also frequently used for switching on additional diagnostics, which
are useful only during debugging sessions.

Another kind of hard-coding in Listing 5.3 was in in *lines 19–20*, where we told
make that all objects files in our project depend on $\boxed{\text{NumericKinds.o}}$ (because
the NumericKinds module selects the precision of most variables used in our
application). However, we need to make $\boxed{\text{NumericKinds.o}}$ an exception to this
rule, to avoid infinite recursion. In Listing 5.3 we reconciled these requirements by
simply listing manually all the other object files. However, in our new version (Listing 5.4), we use an *intrinsic function* of make ($\boxed{\text{filter-out}}$), to construct this list
programatically; all that is required is to construct another list, taking $\boxed{\text{objs}}$ as an
input, and excluding $\boxed{\text{NumericKinds.o}}$—this is exactly what $\boxed{\text{filter-out}}$
does. make has many such intrinsic functions, especially for manipulating strings,

---

[21] make can also use *environment variables*. However, contrary to options specified on the command
line, those have *lower precedence* than variables defined within the makefiles.

working with filenames, etc. Moreover, developers are also allowed to define their own functions.[22]

As a showcase for the last feature of make which we discuss here, we added (*lines 34–35* in Listing 5.4) the $\boxed{\text{clean}}$ target. This target, which can be used as an alternative goal at the command line, removes all of the files that were generated automatically project (i.e. intermediate files with extension $\boxed{\text{.o}}$ or $\boxed{\text{.mod}}$, and the final program executable). In make parlance, $\boxed{\text{clean}}$ is called a *phony target*, since we do not have an actual file with this name—it should be thought of as an "abstract" task. Of course, nothing stops someone from creating a file with this name, which would probably confuse make. To prevent this problem, all *phony targets* should be declared as prerequisites of the special target $\boxed{\text{.PHONY}}$, as we demonstrate on *line 37*; with that syntax, we are clarifying to make that "clear" is not to be treated as a real filename. Phony targets are commonly used by many software packages and, like $\boxed{\text{clean}}$, some have become quasi-standardized—for example, $\boxed{\text{all}}$ (to build all elements in our project), $\boxed{\text{install}}$, or $\boxed{\text{check}}$ (to run any tests that the package may come with, to check proper functioning of the final executables).

Finally, note that in the actual command of the rule for $\boxed{\text{clean}}$, we precede the command by a *minus sign* ($\boxed{-}$). This syntax tells make that it should not abort if this command is not successful. This is necessary in our present case, since it can well happen that at least one of the auto-generated files does not exist in the first place, in which case the removal command will fail, of course.

Exercise 19 (*Using* make *for debugging* makefiles) Modify the makefile in Listing 5.4, so that an extra diagnostic message is printed whenever make matches one of our custom non-phony rules (i.e. the rule on *lines 28–29* or the pattern rule on *lines 31–32*). As part of the diagnostic message, display the name of the current target.
Hint:
variables (normal or automatic) can be passed to the shell, to be displayed (e.g. with the echo-command on Unix).

5.1.2.4 Outlook: Where No **Make** Has Gone Before …

In the pages above, we presented some basic notions about build systems in general (and about make in particular). Here, we provide a short (and very subjective) overview of build system technologies in general (focusing on those which also support Fortran).

---

[22] Indeed, make is a *Turing-complete* language, which means that any imaginable program *could be (in theory)* written in the make language itself—just that it may take a lot more effort than using other languages (which is why make did not make many inroads outside its intended "infrastructure" role).

For small and even medium sized projects, make is a perfectly usable solution, especially when a single development system with some flavor of Unix is used.[23] Gaining some familiarity with make is an excellent way to understand some basic concepts related to build systems. In addition, since many software projects still rely on this tool, it is also a time investment that will pay off throughout a developer's career.

However, the complexity of the make-based solution can quickly increase, as soon as:

- we need some more advanced features (such as separate source and build trees in projects with a nontrivial directory layout), or
- the software needs to compile on multiple machines (with variations in hardware and/or software configurations)

The problem is not even that make-based systems cannot handle the situations above—as we hinted in the preceding sections, *GNU Make* (gmake) in particular is a very powerful tool, which can be (and has been) successfully used to construct systems of arbitrary complexity. However, actually achieving this in practice is a nontrivial task, which is better approached as a distinct software development project on its own, to be handled in parallel to the actual code of the application. Needless to say, this is not a task for novices in make's ways.

As a first approach to some of these problems, many projects began to include a shell script (usually named configure), which performed an analysis of the machine where the software was about to be compiled ("build machine"), and created the makefile, based on the outcomes of this analysis and on a template makefile provided by the package's authors. unix users may have used this command already, which is part of the standard sequence of commands when compiling a package from source[24]:

```
$ ./configure && make && make install
```

Among the tasks usually handled by this script, we have:

- **ensuring that the build machine meets some minimal requirements:** For example, one can check that the necessary program (compiler(s), linker(s), make, shell, file system utilities, etc.) are installed and functional. Also part of this task is ensuring that the necessary libraries are available. If any of these critical checks fails, configure aborts with an error message.

---

[23] Working on Windows is also feasible, especially if some basic GNU! tools are installed (for example, as provided by the Cygwin or MinGW projects). However, developers should be prepared to handle some additional complexity (introduced by using Unix tools into what is essentially a non-Unix environment).

[24] This is different from installing pre-compiled packages through a package manager (such as apt, rpm or yum, employed by many Linux distributions). In general, installing from source is only recommended for software which was not adapted to work with such package managers; unfortunately, many climate models in ESS are in this situation.

- **checking for optional features:** Authors of the software package may want to take advantage of additional technologies when possible, to enable optional features (e.g. advanced visualizations) or to improve performance. The second scenario is common in ESS and *high performance computing* (HPC) in general, since hardware vendors often supply versions of commonly used libraries which are optimized for their systems.
- **modifying the** `makefile`**, to reflect system characteristics:** Once the `configure`-script finalized the analysis of the system, it combines this information with the `makefile`-template, to create a final `makefile`, which is what the `make` program actually uses in the next stage.

Autotools **suite** As the reader may have already guessed, there is nothing easy (or pleasant) about writing the `configure`-script manually. In particular, writing it such that it works correctly across all environments is *really challenging*. Fortunately, developers are nowadays spared this effort, thanks to advanced build systems like the *autotools*[25] suite. Without going into details (see, e.g., Calcote [3] for more on *autotools*), this software suite consists of several programs and libraries (of which `autoconf`, `automake` and `libtool` are most prominent). `autoconf` takes an abstract description (usually from a file named `configure.ac`) of the project's requirements and optional features, and creates impressive `configure`-scripts, which will effortlessly run on most systems. In a somewhat similar fashion, `automake` takes a high-level description (usually from a file named `Makefile.am`) of the `makefile` we want to obtain in the end, and creates a `makefile`-template (named `Makefile.in`). The maintainer of the software package usually provides to users the resulting `configure`-script and the `Makefile.in` file. On their side, users run the `configure`-script, which performs the already mentioned analysis of the build machine and, based on the results and the `Makefile.in`, creates the final `makefile`. The beauty of this system is that users are able to configure and compile the software package, even if they did not install the *autotools* suite—that is only needed on the package maintainer's machine.

While the workflow outlined above is often sufficient, there is an additional component, that readers interested in *autotools* should know about – namely, the creation of the `configure.h` file. This C/C++ header file is helpful for backfeeding information about the build machine into the project's source code; it contains definitions of symbols destined for the preprocessor,[26] which can be used to selectively enable features in the source code.

---

[25] Note that there is no actual program with this name—this is more of an "umbrella" term. Alternatively, this collection of tools is also named the *GNU build system*, because it has become the de-facto build system in the GNU/Linux world.

[26] Most Fortran compilers also allow enabling a C/C++ preprocessor.

The third major component of *autotools* is $\boxed{\texttt{libtool}}$, which hides from the developers the idiosyncrasies of the different platforms with respect to how shared libraries are used.

The "new wave": SCons and CMake In closing our discussion of build systems, we should also mention the "competition" to *autotools*. Noteworthy candidates in this category are CMake and the SCons. While we refrain from giving specific recommendations on which system to use, these alternatives may be worth considering for some of our readers. For Fortran developers, a feature which both SCons and CMake provide (but was notably missing from *autotools* at the time of this writing) is automatic dependency analysis for Fortran code, especially when using modules.[27]

CMake is actually a *meta-build-system*, since it supports multiple *generators*. To understand the difference, *autotools* always create in the end a versatile make-based framework (in a fraction of the time that would be needed if writing the framework from scratch). This works less well with non-Unix platforms (especially in Windows at the time this book was written). CMake is more versatile in this sense, because it also supports creating, for example, native build systems specifications (e.g. Microsoft Visual Studio and OSX XCode projects). In terms of features, there is significant overlap with *autotools*. Also, CMake defines its own programming language, which again implies a learning curve (although the syntax is allegedly friendlier than for makefiles or the shell-scripts-with-macros used by *autotools*).

SCons is another build system that is roughly equivalent to *autotools* in terms of features. Similar to CMake, SCons also has built-in support for non-(Unix) platforms. A primary focus of the system is *build correctness*, which is implemented by also tracking aspects that many other systems miss by default (e.g. changes in include or library paths, or in compiler flags will trigger a re-compilation of the affected object files in SCons—see Smith [26]). However, perhaps the most popular "selling point" of SCons is that it is written as a *domain-specific language* (DSL) embedded within the Python programming language, which makes it very easy to extend, especially for developers which already employ this language for other tasks.

We recommend Smith [26] to readers interested in build systems, for a good overview and comparisons of these technologies. Also, see Martin and Hoffman [16] for CMake, and Knight [12] for SCons (as well as the corresponding websites dedicated to these tools).

5.2 Input/Output

Earlier in Chap. 2, we presented some forms of file-based I/O. Those are, however, inconvenient for nontrivial application (and even more so for large scale modelling in ESS). Notable weaknesses of those simple I/O-techniques are that they are both *not*

---

[27] In principle, this facility is often provided by the compiler and, indeed, it works quite well with C(++) code. However, gfortran had, at the time of this writing, only primitive support for this, which shifts the burden more on the build systems.

self-descriptive (unless programming effort is explicitly allocated to improve this), and also *not space-efficient* (unless the non-portable `binary`-format is chosen). In this section, we describe two established tools to get around these limitations.

5.2.1 Namelist I/O

While reading data from a simple ASCII file (as discussed in Chap. 2), one has to ensure that the values are read into the right variables, and in the right order, to match the contents of the input file. Since there is no easy way to document the data within the file itself, working with such data can become frustrating and error-prone. The concept of *namelist-I/O* in Fortran was designed to help in these scenarios, especially when small amounts of data are involved (e.g. when loading/saving the model parameters in ESS).

There are two components to consider when working with a `namelist`: *namelist groups* (in the Fortran code), and the `.nml` files themselves (where data is stored). We will address both issues below, and afterwards provide a more realistic usage example (by extending the heat diffusion solver from Sect. 4.1).

5.2.1.1 Defining and Working with *namelist Groups*

Namelist groups are defined via statements in the Fortran application. The statements can only appear in the *declarations part* of program units. The general syntax for declaring such a group is[28]:

```
! Declarations for var1, ..., varn
namelist/namelist_group_name/ var1 [, var2, ... , varn ]
! Other declarations ...
! Executable statements of the (sub)program
```

In essence, this tells the compiler that `var1 ... varn` should be treated as a unit in I/O-statements that use this `namelist`. To illustrate, here is how we would define a group which links together two scalar variables (of types `logical` and `real`), an array, and a user-defined type:

```
 8    ! user-defined DT
 9    type GeoLocation
10        real :: mLon, mLat
11    end type GeoLocation
12
13    ! Variable-declarations
14    logical :: flag = .false.
15    integer :: inFileID=0, outFileID=0
16    real :: threshold = 0.8
17    real, dimension(10) :: array = 4.8
18    type(GeoLocation) :: myPos = GeoLocation(8.81, 53.08)
19
```

---

[28] Note that we use the same convention as in earlier chapters, denoting by square brackets any optional elements (i.e. the brackets themselves should not appear in actual code).

```
20    ! namelist-group (binds variables together, for namelist I/O).
21    namelist/my_namelist/ flag, threshold, array, myPos
```

Listing 5.5 | `src/Chapter5/demo_namelist.f90` | (excerpt)

Once the `namelist` has been defined, it can be used in `read`- and `write`-statements. For example, we could write the current program state in a file:

```
26    ! Write current data-values to a namelist-file
27    open(newunit=outFileID, file="demo_namelist_write.nml")
28    write(outFileID, nml=my_namelist)
29    close(outFileID)
```

Listing 5.6 | `src/Chapter5/demo_namelist.f90` | (excerpt)

where in the `write`-statement above we have `nml=my_namelist` instead of the usual format specifier; also, the list of entities to write is missing (the complete `namelist` will be written).

Naturally, reading from a pre-existing namelist file is also possible, allowing us to update some (or all) data in the `namelist` based on that file. For our test program, this looks like:

```
31    ! Update (read) *some* values in the namelist, from another file
32    open(newunit=inFileID, file="demo_namelist_read.nml")
33    read(inFileID, nml=my_namelist)
34    close(inFileID)
```

Listing 5.7 | `src/Chapter5/demo_namelist.f90` | (excerpt)

where a possible input file (created by us with a regular text editor) would be:

```
4    &my_namelist
5       ! Comments can be added on distinct lines...
6       myPos%mLon    = 9.72,      ! ...or at the end of a line.
7       myPos%mLat    = 52.37,
8       array         = 6*9.1,     ! shorthand-notation for constant
9                                  ! sections in an array.
10      array(1)      = 2.9 ! overrides previous specification for
11                              ! first array element
12   /
```

Listing 5.8 | `src/Chapter5/demo_namelist_read.nml` | (excerpt)—a simple namelist file

Note that we can specify components of the `namelist` in any order, and even omit some of these components—these features are summarized below.

Structure of namelist files When creating (or interpreting) a new `namelist` file like the one shown in Listing 5.8, there are several simple syntax rules to consider. First, the ampersand character (`&`) should appear, followed (without any intervening space) by the name of the `namelist` (in our case—my_namelist). After this, the actual information is specified, as *key-value pairs* (such as `var_name = var_` `value`). Each pair can appear on a distinct line, or several of them can be aggregated in a line, separated with commas (`,`). Finally, a slash (`/`) marks the end of the `namelist`-specification.

Some additional observations:

- Throughout the file, it is possible (and even recommended) to write comments as in normal Fortran code, to better document the data entries.
- It is perfectly acceptable to specify only part of the variables in the corresponding `namelist` in such a file. If that is the case, the un-specified variables will not be affected by the `read`-statement. This feature is very convenient for ESS models, since it allows users to write short input files, containing only the parameters they need to change (out of the complete list of model parameters, which can be more intimidating).
- For large arrays, which need to be initialized by a constant value, it is possible to use the shorthand notation `n_repetitions * value`; for example, *line 8* in Listing 5.8 is equivalent to:

```
array = 9.1,   9.1,   9.1,   9.1,   9.1,   9.1,
!          <--------- 6 repetitions --------->
! NOTE:  array(7:10)-elements are not affected.
```

- A variable may be specified more than once. In that case, the specifications can be interpreted as sequential assignments (so the last value will be taken in the end).
- It is not necessary to specify the variables in the same order as they appear in the namelist group definition in the code. The Fortran runtime system will automatically handle the parsing of the file.

The files themselves (often given the .nml-extension) are in human-readable, ASCII format, which is not efficient for large amounts of data (we discuss a solution for that in Sect. 5.2.2).

5.2.1.2 Example: Simplifying the Heat Diffusion Program with Namelists

As a more complex use case for `namelists`, let us consider how we can improve the procedure of reading model parameters for the application discussed in Sect. 4.1. In that version of the code, the parameters were specified in a non-descriptive ASCII file, reads:

```
100.
75.
50.
25.
200
1.15E-6
30.
```

Listing 5.9 `src/Chapter4/config_file_formatted.in` —previous version of input file, for the heat diffusion solver (Chap. 4)

This is not a robust approach, since there is no information (in the file itself) about what each line of input represents. We can easily improve this, by modifying the constructor (= initializer) of the `Config`-type. The changes we need to make (relative to the program `src/Chapter4/solve_heat_diffusion.f90`) are actually minimal, and concentrated in the initializer function (`createConfig`):

```
48   module Config_class
49      use NumericKinds
50      implicit none
51      private
52
53      type, public :: Config
54         real(RK) :: mDiffusivity = 1.15E-6_RK, & ! sandstone
55                    ! NOTE: "physical" units here (Celsius)
56         mTempA = 100._RK, &
57         mTempB =  75._RK, &
58         mTempC =  50._RK, &
59         mTempD =  25._RK, &
60         mSideLength = 30._RK
61      integer(IK) :: mNx = 200 ! # of points for square side-length
62      end type Config
63
64      ! Generic IFACE for user-defined CTOR
65      interface Config
66         module procedure createConfig
67      end interface Config
68
69   contains
70      type(Config) function createConfig( cfgFilePath )
71         character(len=*), intent(in) :: cfgFilePath
72         integer :: cfgFileID
73
74         ! Constant to act as safeguard-marker, allowing us to check if values were
75         ! actually obtained from the NAMELIST.
76         ! NOTE: '-9999' is an integer which can be *exactly* represented in the
77         !       mantissa of single-/double-precision IEEE reals. This means that the
78         !       expression:
79         !          int(aReal, IK) == MISS
80         !       will be TRUE as long as
81         !          (a) 'aReal' was initialized with MISS and
82         !          (b) other instructions (e.g. NAMELIST-I/O here) did not modify the
83         !              value of 'aReal'.
84         integer(IK), parameter :: MISS = -9999
85
86         ! We need local-variables, to mirror the ones in the NAMELIST
87         real :: sideLength=MISS, diffusivity=MISS, &
88            tempA=MISS, tempB=MISS, tempC=MISS, tempD=MISS
89         integer :: nX = MISS
90         ! NAMELIST definition
91         namelist/heat_diffusion_ade_params/ sideLength, diffusivity, nX, &
92            tempA, tempB, tempC, tempD
93
94         open( newunit=cfgFileID, file=trim(cfgFilePath), status='old', action='read' )
95         read(cfgFileID, nml=heat_diffusion_ade_params)
96         close(cfgFileID)
97
98         ! For diagnostics: echo information back to terminal.
99         write(*,'(">> START: Namelist we read <<")')
100        write(*, nml=heat_diffusion_ade_params)
101        write(*,'(">> END: Namelist we read   <<")')
102
103        ! Assimilate data read from NAMELIST into new object's internal state.
104        ! NOTE: Here, we make use of the safeguard-constant, so that default values
105        !       (from the type-definition) are overwritten only if the user provided
106        !       replacement values (in the NAMELIST).
107        if( int(sideLength, IK) /= MISS ) createConfig%mSideLength = sideLength
108        if( int(diffusivity, IK) /= MISS ) createConfig%mDiffusivity = diffusivity
109        if( nX /= MISS ) createConfig%mNx = nX
110        if( int(tempA, IK)  /= MISS ) createConfig%mTempA = tempA
111        if( int(tempB, IK)  /= MISS ) createConfig%mTempB = tempB
112        if( int(tempC, IK)  /= MISS ) createConfig%mTempC = tempC
113        if( int(tempD, IK)  /= MISS ) createConfig%mTempD = tempD
114      end function createConfig
115   end module Config_class
```

Listing 5.10 | `src/Chapter5/solve_heat_diffusion_v2.f90` | (excerpt)

As necessary infrastructure for namelist I/O, we add several local variables (*lines 86–89*), which are packaged into the namelist definition (*lines 90–92*). In *lines 94–96* the namelist is used and, as a debugging facility, the final status of variables in the namelist group is printed on-screen.

The rest of the new code (*lines 74–84* and *103–113*) is necessary to account for the possibility of incomplete namelist files. As already mentioned, this feature is very useful for simplifying interaction with the code. For example, if the user only needs to change the diffusivity of the material (while keeping all other parameters at default values), the input file should contain just the entry for the new diffusivity. To support such partial updates of the configuration, however, we need a mechanism for checking if a parameter's value was actually obtained from the namelist file. Our simple approach here is to initialize all numeric members of the namelist group with a special value (`MISS=−9999`), which is known to reside well outside the valid range for the simulation parameters. Note that `MISS` is an integer, but it can also be used to mark floating-point variables as "dirty" (un-initialized).[29] All local variables will start in this state, and will be transferred as simulation parameters only if updated during the `namelist`-read command (at *line 95*).

As a sample namelist-based input file, we have:

```
 1  &heat_diffusion_ade_params
 2  ! Physical parameters.
 3      diffusivity  = 1.15e-6,  ! thermal-diffusivity coeff (m^2/s)
 4      ! NOTE: commenting-out line below will cause default-value to be picked
 5      sideLength = 10.  ! length of square-side (m)
 6
 7      ! Constant-temperature boundary conditions.
 8      tempA = 100.,
 9      tempB =  75.,
10      tempC =  50.,
11      tempD =  25.,
12
13      ! Numerical parameters.
14      nX = 300
15  /
```

Listing 5.11 | `src/Chapter5/heat_diffusion_config.nml` |–input file for the new version of the heat diffusion solver

By using a `namelist` for specifying our model parameters, the readability of the configuration files has clearly been improved. Since configuration files are often part of the model's "interface" with the users (climate scientists in ESS), it is perhaps not surprising that many ESS models use this technique extensively.

5.2.2 I/O with the NETwork Common Data Format (netCDF)

Although `namelists` are really useful in many cases (e.g. for providing model parameters), they are unsuited for handling larger volumes of data, due to the same types of *storage* and *computing-time* inefficiencies which affect ASCII files[30] (as discussed in Chap. 2). Since large volumes of data are very common in ESS, developers were historically forced to use various forms of binary I/O. However, while such approaches reduce the efficiency problems, they spawned considerable difficulties for scientific collaboration, as most research groups developed their own practices for storing such data, making datasets from different scientists more challenging to compare (on technical grounds) than necessary. Standardization efforts were clearly

---

[29] This works because the absolute value of `MISS` is still small enough to fit into the mantissa of the common floating-point formats. Given our choice of numeric kinds, this ensures that we can compare the integer part of the real variables against `MISS` (*lines 107–108* and *110–113*), without having to worry about floating-point approximation of numbers. **In general, however, note that direct comparisons of real variables should be avoided whenever possible, since this can easily introduce bugs (see also discussion in Sect.** 2.3.2).

[30] Indeed, namelist files *are* ASCII files, just that they require a special format.

necessary (for the benefit of all stakeholders), and the *World Meteorological organization* (WMO) pioneered such work. While those early solutions improved the situation, they still had some technical problems (see, e.g., Caron [4]). In response to these concerns, the *Network Common Data Format* (netCDF) data formats were created, supported by the *University Corporation for Atmospheric Research* (UCAR). In this section, we focus on these latter technologies, which have become the de-facto standard, especially for modelling work in ESS.

In addition to being *platform-* and *language-independent*, the netCDF-formats also *permit efficient I/O*[31] and *creating self-describing datasets.*[32] Another noteworthy aspect is that UCAR aims to keep the software *backwards-compatible* so that, once created, a netCDF-file can still be accessed by future versions of the software.

As high-level components in the netCDF "ecosystem", we can identify:

1. First, we have the *data formats* themselves, which are public specifications of how data is to be stored. Two formats (named **classic** and **64-bit offset**) are also open standards of the *Open Geospatial Consortium* (OGC).

2. In the second layer, we have software libraries (similar to what we described in Sect. 5.1.1), which can read and write data in the netCDF-formats. These are also provided and maintained by UCAR, as a courtesy for application developers.[33] In fact, UCAR provides several such libraries, in two "strands": one for the JAVA-language, and a second strand for compiled native languages. In the second strand we have a C core library, with *Application Programming Interfaces* (APIs) for several languages (C, C++, Fortran 90 and the older Fortran 77). These "wrapper"-libraries depend on the C core library,[34] and so do the many third-party packages which are available for using netCDF within scripting languages (Python, R, IDL, Perl, MATLAB, etc.).

   The use of the common core in the second strand also ensures that programs written in different languages can exchange data via netCDF-files.

3. Finally, in the third layer, we have the applications which use netCDF. Most models in ESS can be included in this broad category, as well as utility packages which facilitate post-processing and plotting of results:

   - **manipulation at the command line**: Readers familiar with Unix will find the cdo and nco packages useful, since they enable powerful manipulations from the command line, and can also be used for automated post-processing (with shell scripts).

---

[31] Depending on the data access patterns of the application, some knowledge about the representation may be necessary for achieving higher performance.

[32] Of course, actually achieving a "self-describing" status is the responsability of the application developers, who know what the data actually stands for—the advantage of netCDF is that it enables *embedding* such information ("metadata") within the same file which holds the binary data.

[33] This is an excellent example of how libraries are useful—in this case, they relieve most scientists from having to worry about how their data is mapped to bits in the computer and the other way around.

[34] The dependency is important, for example, if the libraries need to be compiled from source for some reason.

- **viewers/browsers**: Several application can be used to visualize netCDF-files interactively—for example, $\boxed{\text{ncBrowse}}$, $\boxed{\text{Panoply}}$, $\boxed{\text{nCDF}}$ - $\boxed{\text{Browser}}$, $\boxed{\text{Paraview}}$ and $\boxed{\text{ncview}}$. The latter in particular is very popular for taking a quick look at the data.
- **processing environments**: There are several data processing environments with support for netCDF. General-purpose scripting languages can be classified here, but also $\boxed{\text{NCL}}$, $\boxed{\text{Ferret}}$, $\boxed{\text{GrAds}}$, $\boxed{\text{ArcGIS}}$ and $\boxed{\text{Origin}}$.
- **software libraries**: $\boxed{\text{GDAL}}$ and $\boxed{\text{VTK}}$ can also read netCDF-files.

We can also fit in this category three *Command line Interface* (CLI) utility programs provided by UCAR (usually packaged together with the C library):

- $\boxed{\text{nccopy}}$—for converting between the different file formats supported
- $\boxed{\text{ncdump}}$—for creating an ASCII view of the file; this will usually create *a lot* of output, so it is useful mostly with the $\boxed{\text{-h}}$ flag (which only provides an overview of the file, skipping the data values)
- $\boxed{\text{ncgen}}$—for converting ASCII data to netCDF (opposite to $\boxed{\text{ncdump}}$)

In this section of the book we focus on the interaction between the last two layers listed above. Specifically, we will discuss some basics of using the Fortran 90 *Application Programmimg Interface* API in applications (but omitting features like the newer **netCDF-4/HDF5** data format, *parallel I/O* or *remote data access*).

Versions of *libraries* **and** *binary file formats* **in this book**
Like most successful software projects, netCDF is continuously evolving. Our discussion here is limited to the latest versions available at the time of writing (specifically, $\boxed{\text{Version 4.3.1.1}}$ of the C-core, and $\boxed{\text{Version 4.2}}$ of the Fortran 90 wrapper). The reader is encouraged to consult the official documentation (available from http://www.unidata.ucar.edu/software/netcdf/), for new concepts which may be introduced by later versions, and for definitive information.

5.2.2.1 Overview of Concepts in the NetCDF Data Models

To understand how the various functions in the API fit together, it is useful to have an overview of the high-level concepts in the netCDF data model:

1. **dataset**: In netCDF terminology, a dataset represents the top-level entity, to which variables, dimensions, or attributes belong. In this model, for each dataset we have a corresponding file on the user's computer (which contains, for example, some measurements or model output).

2. **variable**: A variable corresponds to actual data. In `netCDF`, variables are
   n-dimensional arrays of data ($n \geq 0$, where $n = 0$ represents a scalar, $n = 1$
   a vector, $n = 2$ a matrix, etc.). Just like Fortran arrays, these need to have a
   uniform data type. For the `netCDF`-classic format, the data type can be *byte*
   (`NF90_BYTE`), *char* (`NF90_CHAR`), *short* (`NF90_SHORT`), *int* (`NF90_INT`),
   float (`NF90_FLOAT`), *double* (`NF90_DOUBLE`).
   Variables also have a *shape*, which is defined in terms of dimensions (see below).
3. **dimensions**: In the context of structured meshes,[35] we can think of a `netCDF`
   dimension as the set of discrete points along an axis where variables are sampled.
   Each dimension has a *name* (often "lon", "lat", "depth"/"height" and "time" in
   ESS), and a *length* (representing the number of discrete points along the respec-
   tive phase space axis. The length can also be set to `NF90_UNLIMITED`, which
   makes it more flexible. Often in ESS the "time"-dimension is marked as unlimited; this allows to append more data values to the dataset (e.g. when a model
   run is continued, or if more measurements are made at later times). Unlimited
   dimensions are also a point where the specific file format is important: **classic** and
   64bit-offset `netCDF`-files allow at most one such dimension (and only as the *last*
   dimension), while the newer **netCDF-4/HDF5** (also known as the *Common Data
   Model*) has no such limitations. When using unlimited dimensions, the values of
   a variable for a specific index value along the unlimited axis are said to form a
   record.
   As a final point related to dimensions, it is usual to have a $1D$ variable for each
   dimension which is not unlimited. Such *dimension variables* have the same name
   as the corresponding dimension; their role is to specify the discrete points along
   the corresponding axis.
4. **attributes**: These are key-value pairs (where the value is a scalar or a $1D$ array),
   which define *annotations* or other auxiliary information (*metadata*) necessary
   for really making the dataset *self-describing*. Indeed, attributes are a key aspect
   for ensuring that the dataset complies with various conventions (e.g. the *Climate
   forcast* (CF) metadata conventions are recommended in ESS).
   Attributes can be defined as properties of a specific variable (description, units
   of measurement, marker for missing values, reference values for offsets, etc.).
   In addition, we can also have global attributes (belonging to the whole dataset)
   by passing the special tag `NF90_GLOBAL` instead of a variable's *identifier* ID
   when the attribute is defined. For example, we could write as global attributes the
   name and affiliation of the author, date of creation, a reference to an associated
   publication, etc.
5. **groups**: This feature was introduced along with the **netCDF-4** file format, and
   is derived from the same concept in the *Hierarchical Data Format—Version 5*
   (HDF5) library.[36] For brevity, we do not cover this feature. However, to summarize

---

[35] For unstructured meshes, it is possible to define an additional dimension based on the index of
the element/vertex.

[36] In fact, the HDF5 library is a prerequisite when using the **netCDF-4** format, since `netCDF` uses
it internally in that case.

the idea, groups provide a mechanism for organizing the data hierarchically. These are similar in spirit to the Unix directory tree: each dataset has a *root group*, which can have as "children" the usual variables, dimensions, and attributes, as well as other groups (which enables a multi-level hierarchy tree). Each sub-group can be viewed as a separate dataset (with its private variables and attributes), except that dimensions are also visible from children sub-groups.

6. **user-defined types**: Another **netCDF-4** feature (which we also do not cover in detail) is the possibility of defining custom types in addition to the ones permitted in the **classic** format (NF90_INT, NF90_FLOAT, etc.). In principle, such custom types may be useful for storing data which does not fit the usual netCDF model (although they also increase complexity, and may severely restrict the selection of software that can read the data).

5.2.2.2 Versions of the Binary File Format

In addition to different versions of the libraries, developers also need to be aware of the different netCDF *file formats*. We already mentioned these above, but here we provide a quick summary. Currently, the supported formats are (sorted by the time of their appearance) **classic, 64bit-offset**, and **netCDF-4/HDF5**. The first two are perfectly usable,[37] but also have various limitations (especially for *variables* or *records* larger than $2GiB$ with **classic**, or $4GiB$ with **64bit-offset**). The new format, based on HDF5, removes these limitations and adds some new features:

- support for *groups* and *user-defined types*
- support for *parallel I/O*[38]
- support for *data compression*
- *data chunking* (i.e. balancing data layout for multiple access patterns)
- multiple *unlimited dimensions* (made possible by chunking)

We leave these features as topics of further exploration for the interested readers. Nevertheless, since the new format builds upon the previous versions, the material we do cover in the rest of the section should still provide background information relevant for all readers.

5.2.2.3 Using the NetCDF Fortran Application Programming Interface (API)

Object tracking When an application interacts with an I/O-library like netCDF, it needs some mechanism for referring to the various entities in the library. Currently in the Fortran API, this mechanism is based on IDs, which are integer-variables passed as arguments to library functions during invocation. This concept is similar to the unit-number which is automatically assigned by the newunit-feature

---

[37] And are most popular at the time of this writing.

[38] A third-party library (Parallel-netCDF) was developed at *Argonne National Laboratory* (ANL), which allows parallel I/O for the *classic* file format also.

(discussed in Chap. 2). Most entities (the dataset itself, variables, dimensions and attributes[39]) have such IDs, which are tracked internally by the library. The user's interaction with these variables commonly follows one of the patterns below:

- The library returns an ID when a new entity is created, or when the user calls some function from the *inquire*-family (to search for an entity by name, etc.). For example, we get a dataset ID after creating a new dataset with `nf90_create`. Similarly, if we read an existing file and we know that there is a dimension named "lat" in the dataset, we can use the function `nf90_inq_dimid` to retrieve the ID of that dimension.
- Once we acquired an ID-value, we can use this in other library calls, to operate (usually—read, write, or further inquire) on entities. For example, when writing data to a file (with `nf90_put_var`), we need to pass in the previously acquired IDs of the parent dataset and of our variable. The same IDs are needed for the opposite operation of reading data from a pre-existing file (except that we need to call function `nf90_get_var`).

Finally, note that some of the library functions require both input and output IDs as arguments. While the actual values of all IDs are maintained internally by the netCDF-library, users still need to declare and keep track of these variables, which can lead to some complexity. Therefore, it is a good idea to separate the code which interacts with the library from the "core" of the applications. This can be achieved by either grouping the library calls into separate functions (our approach below), or using the new `block` / `end block` construct (not available in all compilers at the time of writing), which allows grouping ID-declarations closer to the library calls that need them.

Error handling When using I/O libraries such as netCDF, there are many points where problems can appear: file system or quota limitations may be reached, *network-attached storage* (NAS) systems may go offline (for cluster users), and sometimes even hard disks may fail. To report such situations to the developer, all functions in the netCDF-library return an *error code*. This mechanism is somewhat similar to *exceptions* in other programming languages (e.g. C++), except that here the program would continue (by default) for as long as possible. It has become common practice to define a *wrapper subroutine*, through which all library calls are made. The version we use here is:

```
! error-checking wrapper for netCDF operations
subroutine ncCheck( status )
  ! ................
  integer(I3B), intent(in) :: status

  if( status /= nf90_noerr ) then
    write(*,'(a)') trim( nf90_strerror(status) )
    stop "ERROR (netCDF-related) See message above for details! Aborting..."
  end if
end subroutine ncCheck
```

Listing 5.12 Wrapper subroutine for calling netCDF functions used in this book

---

[39] Attributes are a special case, since their values are retrieved by name. However, the IDs still exist (denoted as *attribute numbers* in the documentation) and can be useful for writing generic software, which can handle arbitrary netCDF-files.

5.2.2.4 Writing a New NetCDF Dataset

In this section, we outline the steps for creating a netCDF-dataset. After some general considerations, we apply this technique to the RB-LBM code developed in Chap. 4, to significantly improve the I/O efficiency of that program.

To write a new dataset, we first need to create it by calling `nf90_create`. The netCDF-library will then continue to track this dataset internally and allow us to interact with it, until we call the function `nf90_close`. Between these two calls, the dataset is said to be in one of two possible *modes*, as follows[40]:

1. **define-mode**: Immediately after creation with `nf90_create`, the dataset will be in this mode. At this point, the general structure of the dataset (as well as any metadata) needs to be defined. Depending on the specifics of the dataset, this is achieved with a combination of calls to `nf90_put_att` (to define attributes), to `nf90_def_dim` (define dimensions), and to `nf90_def_var` (define variables).

   As mentioned earlier, a convention that is used frequently in this stage is to define, for each dimension, a $1D$ variable with the same name as the dimension. These are also known as *dimension variables*, and provide the one-to-one mappings between discrete indices (i, j, k, etc.) in the variable arrays and actual coordinate values (for example, *longitude*, *latitude*, and *depth/height* in many ESS models, assuming a *structured mesh*).

   To start writing actual data values (including for the dimension variables), we have to specifically instruct the netCDF-library to leave *define-mode* and enter *data-mode*, by calling the function `nf90_enddef`.

2. **data-mode**: The second phase consists of actually writing variable values to our dataset (including dimension variables, if any were declared). The most important function at this time is `nf90_put_var`. This can be used to write either all values in the variable at once or a subset of the variable (lower-dimensional "slice", or even individual scalar value).

   Finally, when there is no more data to be added to our dataset, we signal the end of this stage by closing the file (with the `nf90_close`) function.

Example: adding netCDF-output support for the LBM-MRT solver: Earlier in Chap. 4 we presented an application which solved the *2D Rayleigh-Bénard* (RB) problem, using the *lattice Boltzmann method* (LBM). That initial version of the application could only write results in ASCII files (with an ad-hoc structure, which required a customized parser – see the R script `Chapter4/plotFieldFromAscii.R`).

However, we prepared the ground for improving the I/O of the application, by separating the control logic for writing the output into a base type (`OutputBase`),

---

[40] For brevity, we only describe the most common use case, when these modes are used in a simple linear sequence. However, the netCDF-library also allows switching back and forth between these two modes.

from which the type[41] OutputAscii was derived. In this later type we isolated the portions of the I/O code that were specific to the ASCII format. We now return to this example, to add support for netCDF-output. The natural approach is to define a similar type (we will call it OutputNetcdf), which is also derived from OutputBase. The code is provided below; note that we also split the application into several files, as demonstrated with the box model earlier in this chapter, to make the components of the application easier to understand.

Also, note that all variables with "ID" at the end of their names are initialized by the netCDF-library, during the procedure call where they are first used.

Most of the new code is in the file Chapter5/lbm2d_mrt_rb_v2/Output Netcdf_class.f90, which contains the module OutputNetcdf_class. As usual, in the first part of the module we have the definition of the new type (derived from OutputBase):

```
8    module OutputNetcdf_class
9      use NumericKinds, only : IK, I3B, RK
10     use OutputBase_class
11     use netcdf
12     implicit none
13
14     type, extends(OutputBase) :: OutputNetcdf
15       private
16       ! internal handlers for netCDF objects
17       integer(I3B) :: mNcID, mPressVarID, mUxVarID, mUyVarID, mTempVarID, &
18             mUyMaxVarID, mTimeVarID
19     contains
20       private
21       ! public methods which differ from base-class analogues
22       procedure, public :: init => initOutputNetcdf
23       procedure, public :: writeOutput => writeOutputNetcdf
24       procedure, public :: cleanup => cleanupOutputNetcdf
25       ! internal method
26       procedure prepareFileOutputNetcdf
27     end type OutputNetcdf
28   ! ............... (continues below) ......
```

Listing 5.13 src/Chapter5/lbm2d_mrt_rb_v2/OutputNetcdf_ class.f90 (excerpt)—declarations in the module

Note that we need to use netcdf (*line 11* above), so that the compiler will recognize the netCDF functions which we will invoke later. As internal variables for each instance of our new DT, we have some integers, which keep track of the netCDF-IDs (mNcID is the ID of the file/dataset, and the rest are variable IDs). Also, we bind to the generic interfaces procedures which are specific to this type—these are discussed below.

First, we have initOutputNetcdf:

```
30   ! ............... (continued from above) ......
31   contains
32     subroutine initOutputNetcdf( this, nX, nY, numOutSlices, dxD, dtD, &
33           nItersMax, outFilePrefix, Ra, Pr, maxMach )
34       class(OutputNetcdf), intent(inout) :: this
35       integer(IK), intent(in) :: nX, nY, numOutSlices, nItersMax
36       real(RK), intent(in) :: dxD, dtD, Ra, Pr, maxMach
37       character(len=*), intent(in) :: outFilePrefix
38
39       ! initialize parent-type
40       call this%OutputBase%init( nX, nY, numOutSlices, dxD, dtD, &
41             nItersMax, outFilePrefix, Ra, Pr, maxMach )
42
```

---

[41] Here, "type" is the equivalent of what we would name "class" in C++ or Java.

```
43        if( this%isActive() ) then
44          call this%prepareFileOutputNetcdf()
45        end if
46    end subroutine initOutputNetcdf
47 !  . . . . . . . . . . . . . . (continues below)  . . . . . .
```

Listing 5.14 `src/Chapter5/lbm2d_mrt_rb_v2/OutputNetcdf_`
`class.f90` (excerpt)—initOutputNetcdf procedure

This is the analogue of `initOutputAscii` from the previous chapter. We also call
the "init" subroutine of the underlying base type. However, the actual netCDF-file
initialization is delegated to the subroutine `prepareFileOutputNetcdf`:

```
49  !  . . . . . . . . . . . . . . (continued from above)  . . . . . .
50  subroutine prepareFileOutputNetcdf( this )
51    class(OutputNetcdf), intent(inout) :: this
52
53    ! Variables to store temporary IDs returned by the netCDF library; no need
54    ! to save these, since they are only needed when writing the file-header.
55    ! NOTES: - we have 3 dimension IDs (2D=space + 1D=time)
56    !        - HOWEVER, there is no 'tVarID', since this ID is needed later (to
57    !          append values to this UNLIMITED-axis), so it is stored in the
58    !          internal state of the type (in 'mTimeVarID')
59    integer(I3B) :: dimIDs(3), xDimID, yDimID, tDimID, &
60      xVarID, yVarID
61
62    ! create the netCDF-file (NF90_CLOBBER overwrites file if it already exists,
63    ! while NF90_64BIT_OFFSET enables 64bit-offset mode)
64    call ncCheck( nf90_create( &
65      path=trim(adjustl(this%mOutFilePrefix)) // ".nc", &
66      cmode=ior(NF90_CLOBBER, NF90_64BIT_OFFSET), ncid=this%mNcID) )
67
68    ! global attributes
69    call ncCheck( nf90_put_att(this%mNcID, NF90_GLOBAL, "Conventions","CF-1.6") )
70    call ncCheck( nf90_put_att(this%mNcID, NF90_GLOBAL, SPACE_UNITS_STR, &
71      'channel height $L$") )
72    call ncCheck( nf90_put_att(this%mNcID, NF90_GLOBAL, TIME_UNITS_STR, &
73      'diffusive time-scale $\frac{L^2}{\kappa}$") )
74    call ncCheck( nf90_put_att(this%mNcID, NF90_GLOBAL, PRESS_UNITS_STR, &
75      "$\frac{\rho_0\kappa^2}{L^2}$") )
76    call ncCheck( nf90_put_att(this%mNcID, NF90_GLOBAL, VEL_UNITS_STR, &
77      "$\frac{\kappa}{L}$") )
78    call ncCheck( nf90_put_att(this%mNcID, NF90_GLOBAL, TEMP_UNITS_STR, &
79      'temperature-difference between horizontal walls $\theta_b-\theta_t$") )
80    call ncCheck( nf90_put_att(this%mNcID, NF90_GLOBAL, "Ra", this%mRa) )
81    call ncCheck( nf90_put_att(this%mNcID, NF90_GLOBAL, "Pr", this%mPr) )
82    call ncCheck( nf90_put_att(this%mNcID, NF90_GLOBAL, "maxMach", this%mMaxMach) )
83
84    ! define dimensions (netCDF will return ID for each)
85    call ncCheck( nf90_def_dim(this%mNcID, "x", this%mNx, xDimID) )
86    call ncCheck( nf90_def_dim(this%mNcID, "y", this%mNy, yDimID) )
87    call ncCheck( nf90_def_dim(this%mNcID, "t", NF90_UNLIMITED, tDimID) )
88    ! define coordinates
89    call ncCheck( nf90_def_var(this%mNcID, "x", NF90_REAL, xDimID, xVarID) )
90    call ncCheck( nf90_def_var(this%mNcID, "y", NF90_REAL, yDimID, yVarID) )
91    call ncCheck( nf90_def_var(this%mNcID, "t", NF90_REAL, tDimID, &
92      this%mTimeVarID) )
93    ! assign units-attributes to coordinate vars
94    call ncCheck( nf90_put_att(this%mNcID, xVarID, "units","1") )
95    call ncCheck( nf90_put_att(this%mNcID, xVarID, "long_name", SPACE_UNITS_STR) )
96    call ncCheck( nf90_put_att(this%mNcID, yVarID, "units","1") )
97    call ncCheck( nf90_put_att(this%mNcID, yVarID, "long_name", SPACE_UNITS_STR) )
98    call ncCheck( nf90_put_att(this%mNcID, this%mTimeVarID, "units","1") )
99    call ncCheck( nf90_put_att(this%mNcID, this%mTimeVarID, "long_name", &
100     TIME_UNITS_STR) )
101
102   ! dimIDs-array is used for passing the IDs corresponding to the dimensions
103   ! of the variables
104   dimIDs = [ xDimID, yDimID, tDimID ]
105
106   ! define the variables: to save space, we store most results as NF90_REAL;
107   ! however, for the 'mUyMax'-field, we need NF90_DOUBLE, to distinguish the
108   ! 1st bifurcation in the Rayleigh-Benard system
109   call ncCheck( &
110     nf90_def_var(this%mNcID, "press_diff", NF90_REAL, dimIDs, this%mPressVarID))
111   call ncCheck( &
112     nf90_def_var(this%mNcID, "temp_diff", NF90_REAL, dimIDs, this%mTempVarID))
113   call ncCheck( &
114     nf90_def_var(this%mNcID, "u_x", NF90_REAL, dimIDs, this%mUxVarID))
115   call ncCheck( &
116     nf90_def_var(this%mNcID, "u_y", NF90_REAL, dimIDs, this%mUyVarID))
117   call ncCheck( &
118     nf90_def_var(this%mNcID, "max_u_y", NF90_DOUBLE, tDimID, this%mUyMaxVarID))
119
120   ! assign units-attributes to output-variables
121   call ncCheck( nf90_put_att(this%mNcID, this%mPressVarID, "units","1") )
122   call ncCheck( nf90_put_att(this%mNcID, this%mPressVarID, "long_name", &
123     PRESS_UNITS_STR) )
124   call ncCheck( nf90_put_att(this%mNcID, this%mTempVarID, "units","1") )
```

```
125    call ncCheck( nf90_put_att(this%mNcID, this%mTempVarID, "long_name", &
126         TEMP_UNITS_STR) )
127    call ncCheck( nf90_put_att(this%mNcID, this%mUxVarID, "units","1") )
128    call ncCheck( nf90_put_att(this%mNcID, this%mUxVarID, "long_name", &
129         VEL_UNITS_STR) )
130    call ncCheck( nf90_put_att(this%mNcID, this%mUyVarID, "units","1") )
131    call ncCheck( nf90_put_att(this%mNcID, this%mUyVarID, "long_name", &
132         VEL_UNITS_STR) )
133    call ncCheck( nf90_put_att(this%mNcID, this%mUyMaxVarID, "units","1") )
134    call ncCheck( nf90_put_att(this%mNcID, this%mUyMaxVarID, "long_name", &
135         VEL_UNITS_STR) )
136
137    ! end define-mode (informs netCDF we finished defining metadata)
138    call ncCheck( nf90_enddef(this%mNcID) )
139
140    ! write data (but only for coordinates which are NOT UNLIMITED)
141    call ncCheck( nf90_put_var(this%mNcID, xVarID, this%mXVals) )
142    call ncCheck( nf90_put_var(this%mNcID, yVarID, this%mYVals) )
143  end subroutine prepareFileOutputNetcdf
144  ! ................ (continues below) ......
```

Listing 5.15 | `src/Chapter5/lbm2d_mrt_rb_v2/OutputNetcdf_`

`class.f90` (excerpt)—prepareFileOutputNetcdf procedure

This is where we encounter the first calls to the netCDF-library. To report any errors, we wrap all library calls with the ncCheck-subroutine we already presented in Listing 5.12. After creating the file with nf90_create (*lines 64–66* in Listing 5.15), the dataset enters define-mode. Note that in several parts of the netCDF-library it is possible to combine several options by ┤ior├ing them—we used this technique while creating the dataset, to combine the options NF90_CLOBBER and NF90_64BIT_OFFSET.

In *lines 69–82* we write a few *global attributes* (by passing the flag NF90_GLOBAL to function nf90_put_att), to document what the dataset contains. Afterwards, in *lines 85–87*, we call nf90_def_dim, to define the dimensions for the variables in our dataset. The first two ("x" and "y") are "normal" dimensions, in the sense that their lengths are fixed when the dataset is created (depending on the mesh size calculated from the simulation parameters and the stability/accuracy criteria). On the other hand, the third dimension ("t") is declared as *unlimited*,[42] by specifying the special value NF90_UNLIMITED instead of a length. This allows us, in principle, to re-open the dataset later and append more data to the variables which include this dimension.

In *lines 89–92* we define (by calling nf90_def_var) the three variables corresponding to each dimension. Note that the ID returned for the time variable (this%mTimeVarID) is the only one which will not be lost when the subroutine terminates—the IDs of the space variables are not necessary at later stages, since their values are known already when prepareFileOutputNetcdf is executed (indeed, those variables are written in this procedure, as we shall soon see).

In *lines 94–100*, we write some more attributes (this time—attached to the dimension variables).

Then, in *line 104*, we assemble a 1 D array of dimension IDs, which we use in *lines 109–116*, when we define the variables for the core output field of our simulation (the last variable, however, represents a simple time series, so it only needs tDimID—see *lines 117–118*). As the reader may expect already, we document these variables also, with calls to nf90_put_att (*lines 121–135*).

---

[42] Sometimes, the term *record dimension* is used with the same meaning.

Since there is no more metadata to be written, we end *define-mode* (and enter *data-mode* by calling nf90_enddef on *line 138*. Immediately after that (*lines 141–142*), we use the subroutine nf90_put_var to write the variables for the spatial axes, since they are not dependent on time. Here, the procedure nf90_put_var is used to write all values of the variable arrays at once. However, as we will discuss later, the same procedure also allows writing of single variables, or of subsections of an array—indeed, while the previous library calls dutifully prepared the "context", nf90_put_var takes all the credit, because it is the procedure which actually writes our simulation data on disk.

As the prepareFileOutputNetcdf-subroutine terminates, our dataset will remain in *data-mode*, so that we can later write the time-dependent data (i.e. actual model output and the corresponding time values).

With the dataset prepared by the subroutine prepareFileOutputNetcdf discussed above, it is time to show the subroutine which actually writes the simulation output, as this becomes available during our time sweep. This is the role of writeOutputNetcdf:

```
146   ! .............. (continued from above) ......
147   subroutine writeOutputNetcdf( this, rawMacros, iterNum )
148     class(OutputNetcdf), intent(inout) :: this
149     real(RK), dimension(:, :, 0:), intent(in) :: rawMacros
150     integer(IK), intent(in) :: iterNum
151     ! local variables
152     real(RK) :: currTime
153
154     if( this%isTimeToWrite( iterNum ) ) then
155       ! increment output time-slice if it is time to generate output
156       this%mCurrOutSlice = this%mCurrOutSlice + 1
157
158       ! Evaluate current dimensionless-time (0.5 due to Strang-splitting)
159       if( iterNum == 0 ) then
160         currTime = 0._RK
161       else
162         currTime = (iterNum-0.5_RK)*this%mDtD
163       end if
164       ! append value to UNLIMITED time-dimension
165       call ncCheck( nf90_put_var( this%mNcID, this%mTimeVarID, &
166             values=currTime, start=[this%mCurrOutSlice]) )
167
168       this%mUyMax = maxval( abs(rawMacros(:, :, 1)) )
169
170       ! write data (scaled to dimensionless units) to file
171       ! - dimensionless pressure-difference
172       call ncCheck( nf90_put_var( this%mNcID, this%mPressVarID, &
173             values=rawMacros(:,:,0)*this%mDRhoSolver2PressDimless, &
174             start=[1, 1, this%mCurrOutSlice], count=[this%mNx, this%mNy, 1]) )
175       ! - dimensionless temperature-difference
176       call ncCheck( nf90_put_var( this%mNcID, this%mTempVarID, &
177             values=rawMacros(:,:,3), &
178             start=[1, 1, this%mCurrOutSlice], count=[this%mNx, this%mNy, 1]) )
179       ! - dimensionless Ux
180       call ncCheck( nf90_put_var( this%mNcID, this%mUxVarID, &
181             values=rawMacros(:,:,1)*this%mVelSolver2VelDimless, &
182             start=[1, 1, this%mCurrOutSlice], count=[this%mNx, this%mNy, 1]) )
183       ! - dimensionless Uy
184       call ncCheck( nf90_put_var( this%mNcID, this%mUyVarID, &
185             values=rawMacros(:,:,2)*this%mVelSolver2VelDimless, &
186             start=[1, 1, this%mCurrOutSlice], count=[this%mNx, this%mNy, 1]) )
187       ! - max<Uy> (for bifurcation test criterion)
188       call ncCheck( nf90_put_var( this%mNcID, this%mUyMaxVarID, &
189             values=this%mUyMax*this%mVelSolver2VelDimless, start=[this%mCurrOutSlice]))
190     end if
191   end subroutine writeOutputNetcdf
192   ! .............. (continues below)......
```

Listing 5.16 src/Chapter5/lbm2d_mrt_rb_v2/OutputNetcdf_class.f90 (excerpt)—writeOutputNetcdf-procedure

The subroutine operates primarily with the data array rawMacros (passed from the RBenardSimulation-instance which owns this-instance of OutputNetcdf). This is a 3*D* array, where the first two dimensions are for space, and the third dimension represents the specific LBM "moment" (density/pressure anomaly, horizontal velocity, vertical velocity, or temperature—see discussion in Chap. 4). Then, if the output criterion is satisfied (see `OutputBase`), a new time slice will be written in the dataset. Since the time dimension was declared as unlimited, we evaluate (*lines 159–163* above) the dimensionless time for the specific iteration number (`iterNum`).[43] This is then written to the dataset with a new call to `nf90_put_var` (*lines 165–166*), similar to what we have done in subroutine `prepareFileOutputNetcdf` for the spatial axes. Here, however, we only write a single value, which requires us to specify a value for the `start`-argument (which was optional before). As a value for this argument, we construct inline an array which "wraps" our scalar `mCurrOutSlice` (that tracks the current output slice number).

After calculating our diagnostic variable `mUyMax` in *line 168*, we proceed with writing of the remaining output variables, in *lines 172–189*. Here, we have again calls to `nf90_put_var`. While the last one (*lines 188–189*) is similar to the call we just discussed (for writing `currTime`), the other four calls to `nf90_put_var` are more interesting, as they write the $2D$ arrays from our simulation (representing pressure differences, velocity, and temperature). For these calls, we have to specify the `start`-argument again, only that we pass a $1D$ array with three elements, which indicates to the library what are the coordinates (in index space) of the first value from the data array (specified as the *values*-argument in the same function call). Another new aspect brought by the two-dimensionality of our data arrays (which also applies when writing higher-dimensional data) is the need to specify a value for the `count`-argument, which was optional so far. This new argument informs the library about the sub-region (in index space) for which our data array provides values. Here, this is [`this%mNx`, `this%mNy`, 1], because we are writing $N_x \times N_y$ values (one spatial domain), but for a single time point at a time.

The three subroutines we described above provided the bulk of the functionality for creating the dataset and for adding data to it. Next, we need to make the parent `RBenardSimulation`-instance able to close[44] the dataset, so that the netCDF-library can synchronize the data in its internal buffers with the file on disk (without this step, incomplete/corrupt files may be created!). This functionality is provided by the `cleanupOutputNetcdf`-subroutine below, which calls the `nf90_close` procedure and also instructs the base class instance to free any resources:

---

[43] Note that this also causes the array `mTVals` (which pre-computed the time coordinates for the output slices in `OutputBase`) to become obsolete.

[44] The need for this explicit "tear-down" process may be eliminated if the compiler supports the `final` procedure attribute (Fortran 2008). Using that feature, it would be enough to mark the `cleanupOutputNetcdf` procedure as `final` in *line 24* of Listing 5.13, and the compiler would remember to call this when the `OutputNetcdf` instance goes out of scope.

```
194     ! ............... (continued from above) ......
195     subroutine cleanupOutputNetcdf( this )
196        class(OutputNetcdf), intent(inout) :: this
197        if( this%isActive() ) then
198           call ncCheck( nf90_close(this%mNcID) )
199           call this%OutputBase%cleanup()
200        end if
201     end subroutine cleanupOutputNetcdf
202     ! ..............
203
204     ! error-checking wrapper for netCDF operations
205     subroutine ncCheck( status )
206     ! ..............
207     end subroutine ncCheck
208     end module OutputNetcdf_class
```

Listing 5.17 | src/Chapter5/lbm2d_mrt_rb_v2/OutputNetcdf_

| class.f90 | (excerpt)—subroutines cleanupOutputNetcdf, ncCheck (and end of the OutputNetcdf_class module)

Finally, the last function in the module (but with implementation omitted in the listing above) is ncCheck, which is our wrapper subroutine for error checking.

Next, we need to actually use, in the RBenardSimulation_class-module, the new type of "output sink" presented above. A straightforward approach,[45] which we also use below, is to simply replace previous occurrences of OutputAscii with OutputNetcdf. Specifically, this implies that we now have to use the new module (*line 4* below), and to declare the mOutSink member of the RBenardSimulation type to be of type(OutputNetcdf) (see *line 25* below):

```
1    module RBenardSimulation_class
2       use NumericKinds, only : IK, RK
3       use MrtSolverBoussinesq2D_class
4       use OutputNetcdf_class
5       implicit none
6
7       ! Fixed simulation-parameters
8       real(RK), parameter :: &
9          ! To allow the 1st instability to develop, the aspect-ratio needs to be a
10         ! multiple of $\frac{2 \pi}{k_C}$, where $k_C = 3.117$ (see [Shan1997]).
11         ASPECT_RATIO = 2*2.0158, &
12         ! See [Wang2013] for justification of these parameters.
13         SIGMA_K = 3._RK - sqrt(3._RK), &
14         SIGMA_NU_E = 2._RK * (2._RK*sqrt(3._RK) - 3._RK), &
15         TEMP_COLD_WALL = -0.5, TEMP_HOT_WALL = +0.5
16
17      type :: RBenardSimulation
18         private
19         integer(IK) :: mNx, mNy, & ! lattice size
20            mNumIters1CharTime, mNumItersMax, &
21            mNumOutSlices ! user-setting
22
23         type(MrtSolverBoussinesq2D) :: mSolver ! associated solver...
24         ! NEW (Version 2): Use 'OutputNetcdf' sink instead of 'OutputAscii'
25         type(OutputNetcdf) :: mOutSink ! ...and output-writer
26
27      contains
28         private
29         procedure, public :: init => initRBenardSimulation
30         procedure, public :: run => runRBenardSimulation
31         procedure, public :: cleanup => cleanupRBenardSimulation
32      end type RBenardSimulation
33
34      ! ........................
35   end module RBenardSimulation_class
```

Listing 5.18 | src/Chapter5/lbm2d_mrt_rb_v2/RBenardSimulation_

| class.f90 | (excerpt)—new version, using OutputNetcdf as sink type

---

[45] A more elegant approach would be to allow users to seamlessly switch between the two types of sinks—for example, by adding an optional flag to the function which initializes a RBenardSimulation-instance.

To place some numbers behind our claim for higher performance of the netCDF-format relative to ASCII, on our test machine[46] we found that writing all timesteps for 5 characteristic time intervals (with the rest of the parameters the same as in Listing 4.15), resulted in:

- **ASCII -output**: 7.2Gb of data, written in 488 s (producing over 100,000 files, and writing only the temperature field)
- **netCDF-output**: 6.1Gb of data, written in 110 s (while producing a single file, and writing all output fields, i.e., <u>four times more simulation data</u> than the ASCII version)

After normalizing by the amount of simulation data written, this means that the ASCII version required roughly *5 times more storage space* and more than *17 times more computer time*. While the performance numbers will depend in general on the hardware and on how often output is written, this is a good example of what can be encountered in practice.

5.2.2.5 Reading a NetCDF-Dataset of Known Structure

In this section, we discuss the steps for reading a netCDF-dataset of known structure. This assumption does sacrifice some generality in the interest of keeping the code simple.[47] However, for many programs developed in ESS such a compromise is reasonable. The application which we will later discuss in more depth uses this approach for reading the *World Ocean Atlas 2009* temperature dataset ([14]).

To read data from a pre-existing dataset, we first need to open it by calling the nf90_open-procedure. This is similar to nf90_create discussed earlier, except that the dataset will be set to *data-mode* directly (the *define-mode* is skipped by default). With the dataset opened, we can start reading information stored inside. When the names of the dimensions, variables and attributes inside the dataset are known (as we assume here) this information retrieval process typically consists of two phases: *inquiring for IDs* (based on a known dimension/variable name), and then *retrieving the (meta)data* (based on the previously acquired ID). The specific procedure calls for each of the entities that can appear in a **classic** netCDF-dataset are:

- **dimensions**: Based on the dimension name, the ID of the dimension can be found with a call to nf90_inq_dimid. Then, based on that ID, we can use the procedure nf90_inquire_dimension to determine the length of the dimension.

---

[46] Intel i7 ("Sandy Bridge" generation) CPU, 16Gb RAM, 7200 RPM spinning HDD.

[47] Some of the programs which support the netCDF-formats need to be able to work with any netCDF-files that users may provide (visualization software such as ncview or even the trusty ncdump are good examples here). In such cases, the developers of the software can assume very little about the structure of the input datasets; instead, this information needs to be gathered dynamically at runtime. Note that the netCDF-library also has facilities for this later task, although we do not cover them here.

- **variables**: Similarly, given the name of a variable, the corresponding variable ID can be found via `nf90_inq_varid`. Once the ID is known, the variable (or subsections of it) can be retrieved by calling `nf90_get_var`.
- **attributes**: The attributes are again an exception – they are normally accessed directly by their name (without the need for the intermediate ID). This is achieved by calling the procedure `nf90_get_att`.

Example: reading a netCDF -file from the *World Ocean Atlas 2009* temperature dataset: As a concrete example, we will write a program which reads the mean ocean temperature, as derived from analysis of measurements within the period 01/13/1773–12/25/2008 ([14]), with the goal of calculating the mean seawater temperature profile (as a function of depth).

More specifically, we take as input for our program the file `temperature_annual_1deg.nc`.[48] From a preliminary examination with `ncdump`,[49] we find that our dataset has several dimensions: `lon`, `lat`, `depth`, `time`, and `nv`. We will ignore the last two of these (`time` here has length 1, and `nv` = 2, representing the number of points necessary for specifying an interval along a spatial axis). For the current purpose, we are interested mainly in the variable `t_an` (representing the mean seawater temperature) and in the attribute `_FillValue` of this variable (which marks the points in space where no data was available—e.g., for the locations that were part of the land mask).

For the depth profile, we obviously need to read the depths of each vertical level (stored in the variable `depth` in our dataset). Also, for correctly averaging the temperature at each depth, we need to read geometry information. In this specific dataset, each `t_an` reading represents the estimated mean value of the temperature field, over a rectangular grid cell in lon-lat space. While all cells (at a certain depth) would be identical if plotted in $2D$, their shapes and areas are changed non-uniformly when projected on the sphere. Therefore, in addition to the depth, we need to read the longitude and latitude extents for each cell; these are stored as variables `lon_bnds` and `lat_bnds` in our dataset. Because the grid is uniform in lon-lat space, each of these extents depends only on the corresponding coordinate. For simplicity, we approximate the different vertical levels by homocentric spheres. Denoting by d_i the depth of each level, we have a corresponding sphere with radius $r_i \equiv R_E - d_i$, where we take $R_E = 6.371 \times 10^6$ m as the radius at sea level. With these conventions, the surface area (on the sphere) for each cell i becomes:

---

[48] Available at http://data.nodc.noaa.gov/thredds/fileServer/woa/WOA09/NetCDFdata/temperature_annual_1deg.nc (02/21/2014).

[49] Users of Unix-variants and of the (Cygwin) environment can use `ncdump -h temperature_annual_1deg.nc | less`.

$$S_i = \int\limits_{\lambda_i^W}^{\lambda_i^E} \int\limits_{\phi_i^S}^{\phi_i^N} (R_E - d_i)^2 \cos\phi d\phi d\lambda \tag{5.1}$$

$$= (R_E - d_i)^2 \left(\lambda_i^E - \lambda_i^W\right) \left(\sin\phi_i^N - \sin\phi_i^S\right), \tag{5.2}$$

where $\{\lambda_i^W, \lambda_i^E\}$ and $\{\phi_i^S, \phi_i^N\}$ represent the longitude and latitude extents respectively. It is not difficult to show that the contribution of d_i to the area of a cell is very small ($\sim 10^{-6}\,\%$). This allows us to further simplify the expression for S_i:

$$S_i = R_E^2 \left(\lambda_i^E - \lambda_i^W\right) \left(\sin\phi_i^N - \sin\phi_i^S\right), \tag{5.3}$$

Based on this, we define the mean seawater temperature at level k as the weighted mean:

$$\theta_k = \frac{\sum_i \theta_i S_i}{\sum_i S_i}, \tag{5.4}$$

where the index i runs over the set of all ocean grid cells at level k (the ocean grid cells are those where temperature is not equal to _FillValue).

Even this simple application has several phases (reading data, computing the area of each cell, computing the weighted average, etc.). The *World Ocean Atlas 2009* also contains information for other variables. Therefore, it is worthwhile to make our implementation generic enough to cope with similar datasets (for salinity, dissolved oxygen, etc.—see also Exercise 5.20). We achieve this here by using the *object-oriented programming* (OOP) approach. Most of the code (see file Chapter5/read_noaa_data_netCDF/OceanData_class.f90) is for implementing the data type OceanData, its "init"-function, and the type-bound procedure ("methods"). The basic structure of this module (omitting procedure implementations) is:

```
module OceanData_class
   use netcdf
   use NumericKinds
   use GeomUtils
   implicit none

   type, public :: OceanData
      private
      ! dimension-lengths
      integer :: mNumLon, mNumLat, mNumDepth
      ! arrays to hold data
      real(R_SP), dimension(:), allocatable :: mLonVals, mLatVals, mDepthVals
      real(R_SP), dimension(:,:), allocatable :: mLonBndsVals, mLatBndsVals
      real(R_SP), dimension(:,:,:), allocatable :: mDataVals
      ! additional metadata
      real(R_SP) :: mDataFillValue
   contains
      private
      procedure, public :: getDepths
      procedure, public :: getMeanDepthProfile
      ! internal
      procedure :: cellHasValidData
      procedure :: getCellArea
   end type OceanData

   interface OceanData
```

```
        module procedure newOceanData
    end interface OceanData

contains
    ! .............
end module OceanData_class
```

Listing 5.19 `src/Chapter5/read_noaa_data_netCDF/OceanData_`
`class.f90` (excerpt)—declarations in `OceanData_class`-module

The readers should hopefully feel comfortable with the structure of the application, which is similar to that used in several previous examples (e.g. in Chap. 4). Therefore, here we only discuss in detail the part where the data is read from the netCDF-file. In particular, notice that each instance of our new type `OceanData` encapsulates several data arrays, which need to be filled from the input dataset. This task is performed by our "init"-function `newOceanData`, which creates a new `OceanData`-instance, based on the name of the netCDF-file and on the name of the variable to be read from that file (in our case, those will be "temperature_annual_1deg.nc" and "t_an" respectively). This function is:

```
41  type(OceanData) function newOceanData( fileName, dataFieldName ) result(res)
42      character(len=*), intent(in) :: fileName, dataFieldName
43      ! local vars
44      integer :: ncID, lonDimID, latDimID, depthDimID, lonVarID, latVarID, &
45          depthVarID, dataVarID, lonBndsVarID, latBndsVarID
46
47      call ncCheck( nf90_open(path=fileName, mode=NF90_NOWRITE, ncid=ncID) )
48
49      ! Read-in Dimensions:
50      ! (A) retrieve dimension-IDs
51      call ncCheck( nf90_inq_dimid(ncID, name='lon', dimid=lonDimID) )
52      call ncCheck( nf90_inq_dimid(ncID, name='lat', dimid=latDimID) )
53      call ncCheck( nf90_inq_dimid(ncID, name='depth', dimid=depthDimID) )
54      ! (B) read dimension-lengths
55      call ncCheck( nf90_inquire_dimension(ncID, lonDimID, len=res%mNumLon) )
56      call ncCheck( nf90_inquire_dimension(ncID, latDimID, len=res%mNumLat) )
57      call ncCheck( nf90_inquire_dimension(ncID, depthDimID, len=res%mNumDepth) )
58
59      ! Can allocate memory, now that dimension-lengths are known
60      allocate( res%mLonVals(res%mNumLon), res%mLatVals(res%mNumLat) )
61      allocate( res%mDepthVals(res%mNumDepth), res%mLonBndsVals(2,res%mNumLon) )
62      allocate( res%mLatBndsVals(2,res%mNumLat) )
63      allocate( res%mDataVals(res%mNumLon,res%mNumLat,res%mNumDepth) )
64
65      ! Read-in Dimension-Variables:
66      ! (A) retrieve variable-IDs
67      call ncCheck( nf90_inq_varid(ncID, "lon", lonVarID) )
68      call ncCheck( nf90_inq_varid(ncID, "lat", latVarID) )
69      call ncCheck( nf90_inq_varid(ncID, "depth", depthVarID) )
70      ! (B) read variable-arrays
71      call ncCheck( nf90_get_var(ncID, lonVarID, res%mLonVals) )
72      call ncCheck( nf90_get_var(ncID, latVarID, res%mLatVals) )
73      call ncCheck( nf90_get_var(ncID, depthVarID, res%mDepthVals) )
74
75      ! Read-in Bounds-Variables (for lon/lat)
76      ! (A) retrieve variable-IDs
77      call ncCheck( nf90_inq_varid(ncID, "lon_bnds", lonBndsVarID) )
78      call ncCheck( nf90_inq_varid(ncID, "lat_bnds", latBndsVarID) )
79      ! (B) read variable-arrays (here, 2D arrays)
80      call ncCheck( nf90_get_var(ncID, lonBndsVarID, res%mLonBndsVals) )
81      call ncCheck( nf90_get_var(ncID, latBndsVarID, res%mLatBndsVals) )
82
83      ! Read-in data-field-Variable (and associated attribute "_FillValue")
84      call ncCheck( nf90_inq_varid(ncID, trim(adjustl(dataFieldName)), dataVarID) )
85      call ncCheck( nf90_get_att(ncID, dataVarID, "_FillValue", &
86          res%mDataFillValue) )
87      call ncCheck( nf90_get_var(ncID, dataVarID, res%mDataVals) )
88
89      call ncCheck( nf90_close(ncID) )
90  end function newOceanData
```

Listing 5.20 `src/Chapter5/read_noaa_data_netCDF/OceanData_`
`class.f90` (excerpt)—"init"-function

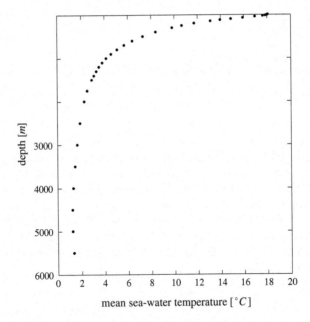

depth [*m*]

mean sea-water temperature [$^{\circ}C$]

Fig. 5.1 Depth profile of seawater temperature, obtained by averaging in time (annual) and in space (at each depth level) of the [14] dataset

For our purposes here (demonstrating how to read a netCDF-file), the interesting code begins at *line 47*, where the dataset is opened. By choosing NF90_NOWRITE as the mode-argument, we protect the file from accidental overwriting of data (possible alternative modes are NF90_WRITE, for appending data to existing datasets or NF90_SHARE for allowing data to be read by a process while another process is writing[50]).

In *lines 51–53*, we obtain the IDs of the dimensions relevant to our task. These IDs are then used in *lines 55–57*, where the sizes of the dimensions are read. After preparing the arrays which will hold our data (*lines 60–63*), we proceed with reading the variables containing the dimension information. This again involves a two-step approach, whereby the functions nf90_inq_varid and nf90_get_var are called for each variable (*lines 66–73*). The exact same approach is used (*lines 76–81*) to read the bounds of the cells and, in *lines 84–87*, to finally read the temperature field. The only peculiarity for this last operation is that for the temperature field (t_an) we also need to read the special attribute which documents missing values (*lines 85–86*).

Since no additional information needs to be read, we can close the file (*line 89*). The curious reader will find a plot of the extracted temperature profile in Fig. 5.1.

---

[50] Modes can also be combined (when this makes sense) using the ior-function, as discussed in Listing 5.15.

Exercise 20 (*Extracting the salinity profile*) Modify the code for reading the *World Ocean Atlas 2009* dataset (see directory `Chapter5/read_noaa_data_netCDF`), to extract the salinity profile instead.
Hint:
for this task, it should be sufficient to modify the file `Chapter5/read_noaa_data_netCDF/read_noaa_woa_2009_data.f90`, which contains the main-program of the application. The required dataset (`salinity_annual_1deg.nc`) can also be found at the same location as the temperature data (http://data.nodc.noaa.gov/thredds/fileServer/woa/WOA09/NetCDFdata/temperature_annual_1deg.nc, as of 28.02.2014).

Exercise 21 (*Heatmap of the temperature profile*) The temperature profile extracted in Fig. 5.1 corresponds to an average over the entire globe. Obviously, this operation filters out all geometric information—a less drastic reduction of the dataset would be more interesting, whereby only zonal averaging is performed. Implement the necessary type-bound procedures to support this operation in module `OceanData_class`. Using your visualization tool of choice, plot the resulting matrix as a heatmap, with depth as the x-coordinate and latitude as the y-coordinate.

5.3 A Taste of Parallelization

Similar to the examples in Chap. 4, many models in ESS rely on discretizations of space and time, to approximate numerically the derivatives in the governing equations. The quality of the approximations generally improves if a finer mesh is used. Unfortunately, mesh refinement also increases the computational time and storage requirements of models (in $3D$ problems, using two times more points along each coordinate leads to an eight times increase in memory used[51]). Due to this polynomial scaling, many models (especially for frontline research in ESS) can easily saturate the computing capacity of a single processor. It is therefore fitting for us to provide at least a brief introduction here. After a high-level overview of trends in the computing industry and of corresponding software technologies (Sect. 5.3.1), we focus on OpenMP, which is a parallel programming model that is widely available

---

[51] This estimate assumes *uniform* refinement. Some models also use *adaptive* mesh refinement, which is often more economic.

and also relatively easy to learn for beginners (Sect. 5.3.4). Finally, in Sect. 5.3.5, we apply OpenMP to some of the example applications from Chap. 4, to demonstrate this technology "in vivo" for the readers.

5.3.1 Parallel Hardware Everywhere …

An "obvious" solution (at least *in theory*) for improving performance is to use multiple processors, which share the work of the application. Indeed, especially for large supercomputers, this approach has become standard practice since the early days. However, until approximately 2002, parallelism was less common outside these large facilities, and relatively few developers used such technologies regularly. One reason for this limited adoption was that parallel programming certainly has a learning curve; for example, many types of bugs[52] which are not possible in the "serial world" can appear. In addition, parallel programs are often more difficult to develop and understand, relative to their serial counterparts. Another major reason was that hardware manufacturers managed, with each new generation of machines, to significantly improve performance even for serial programs. This "free" speedup was good enough for many developers, who could simply rely on the next hardware upgrade cycle to improve the performance of their applications. Withoug delving into details of computer architecture, we can distinguish several broad classes of hardware innovations, which supported these performance increases:

1. *increasing CPU clock rates*: Hardware vendors have continued to find new ways to increase the operating frequency of the CPUs. Assuming, for simplicity, that each instruction (e.g. integer addition) takes a constant number of CPU cycles, decreasing the duration of each cycle would *probably* increase the number of instructions that can be completed in a given time. However, this "frequency race" had to stop [27], as power limitations were reached.
2. *CPU architecture advances*: Unbeknownst to many programmers, parallelism has long been present at the hardware level, even for "single core" CPUs. The underlying idea is to re-arrange and group the instructions of the serial application, such that at least some of the work is parallelized, while still preserving the comfortable *illusion* of serial execution for the developers. These techniques are known as *instruction-level parallesim* (ILP); examples in this class are instruction pipelining, out-of-order execution, small-scale vectorization,[53] etc.

---

[52] Here, we can distinguish between *correctness bugs* (the program produces false results) and *performance bugs* (the program is not using the resources of the underlying hardware efficiently).

[53] The main idea here is to have instructions which operate on arrays instead of on scalar values. Here, by vectorization we mean the *single instruction, multiple data* (SIMD) units of modern CPUs. The term "vector computer" is also used, to refer to systems which implement the SIMD idea on a much larger scale (see, e.g., Hager and Wellein [8] for details). While these machines have many features which make them attractive for scientific computing, they became a niche product by the time of our writing. Instead, hardware with vector-like capabilities, such as the SIMD units and *general-purpose graphics processing units* (GPGPUs), are becoming increasingly popular.

Unfortunately, while it would be convenient to continue improving the performance of CPUs for existing serial code, continuing to do so would lead to poor energy efficiency [19].

Caches and the memory hierarchy

Interestingly, even if faster CPUs were to be created somehow, it would turn out that current memory technologies would probably be unable to feed them fast enough with instructions and data to operate on. The cause is that "raw" memory performance increased at much slower rates compared to CPU-performance [19]. There are two main metrics for performance of memory (or of any communication mechanism, in general):

1. *bandwidth*—how much data can be transported from/to memory per unit time, and
2. *latency*—how large is the time delay, from the moment some data was requested by the CPU until when data starts arriving from main memory.

The problem is that, *if measured in terms of clock cycles*, memory latency actually degraded. Unaddressed, this could create a serious performance problem, since many applications would become memory-bound. Fortunately, hardware vendors mitigated this effect, by including several (currently three) layers of higher performance *cache memory* between the CPU and main memory. As the reader may expect, cache memory is also very expensive, so only small amounts may be added in a system. Fortunately, by sensible management of this limited resource, it is often possible to serve most data accesses from cache. The effect is the best of both worlds: accesses mostly at the speed of the cache, with the capacity of main memory.

The CPU and/or the compiler will typically strive to maximize the usefulness of the cache, by pre-fetching information before it is necessary. However, the success of this enterprise depends ultimately on the code that needs to be executed. This is a point where programmers can make a difference, by writing code which is "cache-friendly", with good spatial and temporal locality. For more details on the memory hierarchy, and on how to optimize code from this perspective, we refer to the book of Hager and Wellein [8].

With the "bad news" out of the way, a positive aspect is that vendors still manage to increase the number of transistors that can be placed on a chip; this is also known as Moore's "law" [22]. These additional transistors are nowadays used to support *explicit parallelism* at all levels, from consumer hardware to supercomputers—serial computers have become the exception. Given these tendencies and considering, in addition, that computational demands in ESS (and other fields) are likely to continue increasing in the future, most scientific programmers need to add parallelization to their skills.

5.3.2 Calibrating Expectations for Parallelization

There are several plausible reasons for considering parallelism. For example, we might be interested in minimizing the time to solution, being able to solve larger problems, increasing throughput, or decreasing the power expended for achieving a result.

For simplicity, we focus here mostly on the first goal (*minimizing time to solution*), where multiple execution units are made to work in parallel, to solve *a problem of constant size* faster. Note that very often the second goal (*solving a larger problem*) is also quite common (for example, when switching to a higher-resolution grid in a ESS model).

When the size of the problem is kept fixed, we can define *speedup* (also known as "scalability") as a simple metric for the effectiveness of a parallelized program:

$$S(N) = \frac{T_1}{T_N} \tag{5.5}$$

where T_1 represents the necessary computing time when using a single execution unit (e.g. single core), and T_N is the time when using N execution units. Ideally, we would have *linear speedup* (also known as "perfect speedup" or "perfect scaling"):

$$S(N)^{\text{ideal}} = N \tag{5.6}$$

However, real-world speedup is often less.[54] To quantify how much less, *parallel efficiency* is commonly calculated, as:

$$\epsilon(N) = \frac{S(N)}{S(N)^{\text{ideal}}} = \frac{S(N)}{N} \tag{5.7}$$

Ideal conditions are then corresponding to $\epsilon = 1 \equiv 100\,\%$.

5.3.2.1 Idealized Performance Models for a Non-ideal World

There are multiple reasons why good speedup may not be achievable. A first such reason is that not all work in a program may be parallelizable. For example, in an ESS model, some model parameters may need to be read at the beginning of a simulation, prior to any calculation. Consider Fig. 5.2a, where we sketch these different types of workloads for an application running on a single processor/core. Let us denote

---

[54] Interestingly, it is also possible (in rare cases) to get *superlinear speedup*, where $S(N) > S(N)^{\text{ideal}}$. This can happen, for example, if a problem is too large to fit inside the cache of one processor, but small enough to fit into the aggregated caches of the N processors.

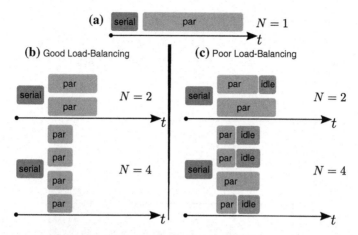

Fig. 5.2 Simplified scenarios for division of work in parallelization: **a** initial serial application, with some "parallelizable" work, **b** parallel execution with good load balancing, and **c** parallel execution with unbalanced workloads (some processors spend significant amounts of time waiting for latecomers). N represents the number of processing units (cores)

by T_1^s the time spent on the non-parallelizable tasks[55] of the program (labeled as "serial" in the diagram), and by T_1^p the remaining time, spent on tasks which *could be* parallelized (labeled as "par" in the diagram); by definition, we have:

$$T_1 = T_1^s + T_1^p \tag{5.8}$$

It is important to notice that, when we use $N \geq 1$ cores/processors, we can only accelerate the parallelizable part (from T_1^p to $T_N^p < T_1^p$). However, the serial part of the workload cannot be reduced; this can introduce a hard lower bound to the execution time, no matter how many execution units are used.

As a first example, consider Fig. 5.2b, where we illustrate the most optimistic scenario, which assumes that (1) the parallelizable work can always be divided evenly between the available execution units and that (2) there is no communication/management overhead associated with the parallel subtasks. The total execution time for our application would be:

$$T_N^o = T_1^s + \frac{T_1^p}{N} \tag{5.9}$$

For brevity, we denote by f_s the fraction of serial work:

$$f_s \equiv \frac{T_1^s}{T_1} \tag{5.10}$$

---

[55] For simplicity, in our sketch we placed the serial fraction at the beginning of the program's runtime. However, periods of serial execution ("serialization") are often distributed more widely throughout the runtime of the program.

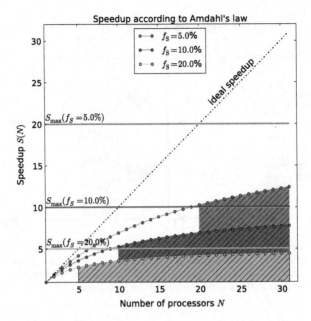

Fig. 5.3 Parallel speedup, as predicted by Amdahl's law, when no load inbalance occurs. For illustration, we present three values of the serial work fraction: *a* $f_S = 5\%$ (*green*), *b* $f_S = 10\%$ (*red*), and *c* $f_S = 20\%$ (*cyan*). For each curve, we show the range of processor counts where efficiency drops below 50 % (*hatched area*), and the maximum achievable speedup S_{max} (*continuous line*)

This allows us to express the optimistic speedup as:

$$S^o(N) = \frac{T_1}{T_N^o} = \frac{1}{f_s + \frac{1-f_s}{N}} \qquad (5.11)$$

which is also known as *Amdahl's law* [1]. We illustrate the predicted speedup in Fig. 5.3. Note that, even with our optimistic assumption of ideal load balance, there is an upper limit on the achievable speedup ($S_{max} = 1/f_S$). Also, for $f_S \ll 1$, parallel efficiency drops below 50 % as soon as we achieve half of the maximum speedup.

The expected speedup is even worse if the work inside the parallel region cannot be distributed equally among the processors (see Fig. 5.2c). Fortunately, for many applications it is more useful to *increase the problem size* (while keeping the amount of work per processor roughly constant). This scenario, also known as *weak scaling*,[56] leads to much more encouraging speedup numbers. We recommend the books of McCool et al. [19] and of Hager and Wellein [8] for more advanced performance models, which consider the weak scaling scenario, as well as other important factors

---

[56] The situation for Amdahl's law, where the total problem size is kept constant, is known as *strong scaling*.

(load inbalance, communication overhead, etc.). Such simplified models of application performance are useful to keep in mind when evaluating a parallel application. While the theory cannot account for all interactions between all software and hardware on a machine, it can be used as a *complementary* tool to real-world performance (measured through *profiling*)—significant disagreement between theory and reality can accelerate identification of performance bottlenecks.

5.3.3 Software Technologies for Parallelism

As already mentioned in Sect. 5.3.1, it is nowadays mostly the responsibility of programmers to identify opportunities for parallelization. Sometimes, the algorithms may also need to be altered, so that the concurrency in the application can be mapped better onto the underlying hardware. As a final stage, these ideas have to be expressed in actual software. To facilitate this last task, several *parallel programming models* were created (and new ones continue to appear). Initially, these were mostly vendor specific extensions of serial programming languages. However, as software developers and users became increasingly concerned with portability of software (to avoid vendor "lock-in"), many of these technologies aggregated into open standards. To the best of our knowledge, the most popular such open standards in the ESS community are currently *Open MultiProcessing* (OpenMP), *Message Passing Interface* (MPI), and *Open Computing Language* (OpenCL), which we summarize below.

- OpenMP is useful for writing software for *shared memory* parallel machines (which also includes most commodity personal computers nowadays, as well as individual nodes in larger supercomputers). The assumed machine model consists of multiple processors which share access to a common memory. This model makes data sharing between processors/cores very easy, which can be both *good* (implicit communication) and *bad* (synchronization problems can be easily introduced, which may take effort to debug).[57] Another characteristic of OpenMP is that it is used mostly through compiler pragmas (comments with special syntax, describing the parallelism in the application), which brings some benefits (discussed later). OpenMP supports Fortran, C, and C++.

- MPI is useful for systems with a *distributed memory*[58] topology. This platforms generally considers each processor as connected to its private memory area—any communication is achieved through explicit messages. Due to this more verbose communication, MPI is in some sense "lower-level" than OpenMP. However, this can also improve scalability to much larger numbers of cores on supercomputers. MPI supports Fortran and C.

- Co-Array Fortran This being a Fortran book, it is fitting that we should at least mention *Co-Array Fortran* (CAF). This is a new set of language extensions (intro-

---

[57] This is the source of most complexity in OpenMP. In general, communication (implicit or explicit) is the point where all parallelization technologies claim the attention of the programmer.

[58] Note that MPI can also be used for *shared memory* machines.

duced by the Fortran 2008 standard), which provide native support for paralleliza-
tion in Fortran. This belongs to the class of languages known as *Partitioned Global
Address Space* (PGAS), which combine aspects of both MPI and OpenMP, with
very concise semantics. Despite being a very interesting new language feature in
Fortran, it is beyond the scope of our text (interested readers can consult, e.g.,
Metcalf et al. [21] for more information).

- OpenCL and OpenACC are newer standards, catering for the increasing popularity of GPGPU and other compute-accelerators such as the Intel Xeon Phi. OpenCL
  is implemented as a C/C++-language dialect, while OpenACC can be viewed as
  a set of pragmas (compatible with C, C++, as well as Fortran), similar in spirit to
  OpenMP.

Interestingly, many HPC applications today use a *hybrid* approach to parallelization,
combining two or more of the parallel programming models above. The boundaries
between these models are also becoming less distinct as the standards are evolving;
for example, the more recent versions of OpenMP (4.0) also introduced support
for SIMD vectorization and for compute-accelerators such as GPGPUs. We do not
cover these features in this text, but interested readers may want to keep an eye on this
technology as compiler support matures, since it could provide a unified platform
for all types of parallelism within a node.

5.3.4 Introduction to Open MultiProcessing (OpenMP)

Out of the large set of parallelization technologies, we elaborate more only on
OpenMP, which is a popular choice for shared memory systems, where the different computational units ("cores") have the same view of the locations in memory
(shared address space).

> **Version of OpenMP in this book**
> Unless specified otherwise, we describe here a (subjective) subset of the
> Version 3.1 of the OpenMP-standard. In Sect. 5.3.5.4, we provide more
> details of the features we *do not* cover here, and provide pointers to relevant
> literature.

5.3.4.1 Basic Syntax and Usage of OpenMP

Our approach for illustrating OpenMP-concepts will depend on code examples. Here,
we summarize some of the "infrastructure" provided by OpenMP, to make it easier
for the reader to follow these examples.

The purpose of OpenMP is to augment a serial programming language (Fortran, C, or C++) by adding support for parallelism. This is achieved with *directives*, *runtime library routines* and support for a set of *environment variables*.

Enabling/disabling OpenMP As a first step, we need to ensure that OpenMP-support is enabled. With most compilers,[59] it is necessary to do this explicitly, with flags that are added to the compilation and linking commands (e.g. $\boxed{\texttt{-fopenmp}}$ for gfortran or $\boxed{\texttt{-openmp}}$ for ifort).[60] For example, using gfortran on a Unix system, we can easily re-compile our first Fortran program (Sect. 2.1) with OpenMP-support[61]:

```
$ gfortran -o hello_world hello_world.f90 -fopenmp
```

Because we did not include any OpenMP-directives in the source code, there will be no noticeable change in the outcome (i.e. it will print $\boxed{\text{"Hello, world of Modern}}$ $\boxed{\textit{Fortran!"}}$ once); soon we will change that.

Note that when (unlike the example above) the compilation and linking stages are separated we need to specify the $\boxed{\texttt{-fopenmp}}$ flag for each phase: at compile-time it causes the compiler to interpret the OpenMP-directives, while at link-time it adds the OpenMP-runtime to the list of libraries that are "linked-in" inside the executable.

Sometimes we may want to disable OpenMP-support, and temporarily revert[62] to the serial (single-threaded) execution, to help debugging or validating the parallel program. When the project contains more than one source code file, it is generally a good idea to support easy toggling of OpenMP-support with a simple switch passed to the build system, to increase productivity when switching back and forth.

The idea of being able to run a program either serially or in parallel, while expecting correct[63] results in either case, is known as *sequential equivalence*; one of the core design principles for OpenMP was precisely to enable developers to write such

---

[59] Cray compilers, which enable OpenMP by default, are an exception.

[60] Please check the documentation of your compiler for exact information about the flags to use, if any.

[61] Readers using the bash shell may use the brace-expansion mechanism, to avoid having to type names repeatedly; for example, we would use: $\boxed{\texttt{gfortran -o hello\_world \{,.f90\}}}$ $\boxed{\texttt{-fopenmp}}$. Other shells may offer similar features.

[62] Some applications are inherently multi-threaded, so a serial version may not make sense. However, this is not the case for many applications in ESS.

[63] Note that the result of the serial and parallel runs *may not be bit-identical*. This happens when floating-point calculations are involved, because such operations are not associative. Due to this missing property, it is quite common with OpenMP to encounter differences (usually in the last digits) in floating-point results, when comparing serial and parallel runs (or even between different realizations of a parallel run), because the non-deterministic scheduling of threads may cause partial results to be accumulated in different orders. In a sense, all results are "correct", so it may be better to accept a range of results during the validation phase. If this is not acceptable (i.e. results need to be strictly reproducible), it is also possible to add ordering constructs in OpenMP; however, doing so will probably decrease parallel performance.

software. A second major principle is *incremental parallelism*, whereby a serial application is transformed into a parallel one incrementally, by identifying (with profiling tools) time-consuming portions of the code, parallelizing those portions, then profiling again, etc. We will use this approach for our parallelization case studies, in Sect. 5.3.5.

Directives As mentioned already, it is the responsibility of the programmer to express parallelism within the application.[64] Parallelism is specified via *directives* which, in the case of Fortran, are written inside what are known as *structured comments*. While to the normal compiler these look like comments (to be discarded without much hesitation), a compiler with OpenMP enabled will pay attention to the directives and generate parallel code accordingly. For modern Fortran, the basic syntax for these structured comments is:

```
!$omp <DIRECTIVE_NAME> [OPTIONAL_CLAUSES]
    !  . . . (block of code affected by directives) . . .
!$omp end <DIRECTIVE_NAME> [OPTIONAL_END_CLAUSES]
```

The first part, `!$omp`, is known as a "sentinel": it signals to the compiler that what comes after the whitespace should be interpreted according to the rules of OpenMP. Loosely speaking, the directives specified after this sentinel orchestrate the parallel execution flow in the program (forking/joining of threads, synchronization, etc.). Most directives also support optional *clauses* for fine-tuning. Within our examples for this section, we will also present some such clauses, e.g., for requesting a specific number of threads, for controlling the way variables are shared between threads, or for changing how thead-local variables relate to corresponding values outside the parallel regions. Note that in Fortran the directives should also be closed, to mark the end of the block[65] of code affected by the directive. The syntax is similar to the normal flow-control constructs (e.g. do−enddo). The block of code surrounded by the directives should be "well-behaved", especially with respect to "jumps": execution should start at the top, and finish at the bottom (no midway-jumping to/from portions of code outside the block is allowed).

Continuing directives Similar to normal Fortran source code, an OpenMP-directive can be continued on the following line, by appending a `&`-character at the end of the current line. This feature is particularly useful when the list of clauses is long.

Conditional compilation As part of the OpenMP-specification, there is also a library of procedures, which can be invoked at runtime, to inquire or modify various aspects of parallelization, or for measuring execution time for sections of code. However, when an application uses procedures from this library, special precautions are necessary when sequential equivalence is to be preserved: somehow, these procedure calls

---

[64] OpenMP only aims to make this process more palatable than working directly with low-level OS threading libraries.

[65] This is not necessary for C and C++, which use curly brackets to surround blocks of code.

need to be removed when compiling a single-threaded (serial) version of the application (otherwise, the application will fail to build properly). For Fortran programmers, two techniques are available to solve this issue:

1. *the* `!$` *sentinel*: For situations when we just need to remove code from a serial build (but there is no code that needs to be enabled *only for serial builds*), we can use the special sentinel `!$`, which is part of the OpenMP-specification. A common use of this technique is to conditionally include the line where the OpenMP-library's module is used:

```
!$ use omp_lib
```

Here, the sentinel is replaced by whitespace if OpenMP-support was enabled at compile-time, and the compiler will treat the line as a normal use-statement. If, on the other hand, OpenMP-support was disabled (or is not present, e.g., for older compilers), the entire line becomes a normal comment (which is discarded).

2. *using the preprocessor*: Very often an "else"-branch is also needed, i.e., to have some code included only for *single-threaded builds* (when OpenMP is disabled). For example, to measure the execution time of a block of code, we want to use `cpu_time` for single-threaded applications, but `omp_get_wtime` when we have multiple threads. In such cases, the usual practice is to use what is known as a *preprocessor*. For our purposes here, a preprocessor can be regarded as a software tool, which can perform some simple transformations on source code (string substitutions, conditional inclusion of lines of code, etc.), passing the result to the compiler. The preprocessing step is usually activated with a compile-time flag (`-cpp` for gfortran and `-fpp` for ifort). For example, we could use the following piece of code to select the appropriate procedure for measuring execution time:

```
1  #ifdef _OPENMP
2     currTime = omp_get_wtime()  ! parallel
3  #else
4     call cpu_time(time=currTime)  ! serial
5  #endif
```

Note that lines of code intended for the preprocessor start with the hash (also known as "sharp") symbol `#`. A compiler with OpenMP-support activated will define the `_OPENMP` symbol, so the ifdef (which stands for "if defined") will evaluate to "true" in that case, causing *line 2* above to be kept and *line 4* to be removed from the code that will be analyzed by the compiler (the preprocessor pragmas will also be removed).

5.3.4.2 Execution Model: Serial and Parallel Regions

Assuming for simplicity that there is no load imbalance within the parallel regions, the basic execution model for a program using OpenMP can be viewed as a generalization of the scenario we used for deriving Amdahl's law (in Sect. 5.3.2.1). The main

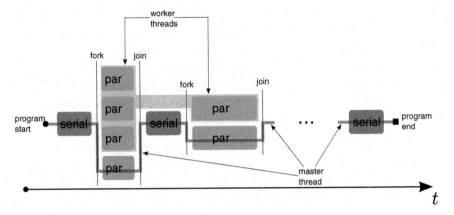

Fig. 5.4 Schematic of the OpenMP execution model. The *gray line* represents the master thread. During the serial sections in the program, only the master thread is working. However, when a parallel section is encountered, a team of threads is forked, which work together with the master until the next serial section is encountered. Threads are given integer IDs *starting at zero*, with ID 0 assigned to the master thread

difference is that, whereas earlier we assumed that the non-parallelizable work is grouped in a single block, we now have *alternating serial and parallel regions* (see Fig. 5.4).

To use more exact terminology, when we launch an application the OS creates a *process*, which is a software entity that groups the instructions and data associated to the application's instance. Within the process, entities known as *threads* are created. Threads are more lightweight units of execution compared to processes, and can be dynamically created and destroyed (supporting the fork-join model). When the application starts, a single thread is running (the "master" thread). However, as soon as a parallel region is encountered (Fig. 5.4) more threads ("workers") are created ("forked", "spawned"), to cooperate with the master on the parallel workload. When this work is finished, the workers "join" with the master-thread. The master then continues execution alone, until the next parallel region is encountered, when the fork/join-process repeats, etc.

To give an example, this is how we could parallelize the "hello world" program:

```fortran
program hello_world_par1
  implicit none

  !$omp parallel
    print*,"Hello, world of Modern Fortran! Parallel too!"
  !$omp end parallel
end program hello_world_par1
```

Listing 5.21 src/Chapter5/hello_world_par1.f90

In *line 4* we specify the parallel-directive, which creates a team of threads. Each of the threads will execute the following block of code, up to the line which closes the directive (*line 7*). In our case, the block of code actually consists of a single line (the `print`-statement, on *line 6*). When run, the program above will print our message several times.

How many threads are there? Exactly *how many times* the message will be printed on your system (i.e. how many threads will be in the team) depends on several factors. Usually, the OpenMP runtime library will take this number equal to the number of (logical) cores in the system. However, *programmers may request a different number of threads*, by specifying a value to the $\boxed{\text{OMP\_NUM\_THREADS}}$ environment variable. For example, on `Unix` systems using the `bash` shell, we may use:

```
$ OMP_NUM_THREADS=2 ./hello_world_par1
```

to request two threads just for one run, or export the value, to set the requested number of threads globally (will apply to all programs started from that shell instance):

```
$ export OMP_NUM_THREADS=2
```

The advantage of the environment variable is that users have the freedom to decide how many threads to use. However, it is also possible to specify the number of threads in the code itself, by adding a `num_threads`-clause to the `parallel`-directive. For example, the following programs asks the user to specify a number of threads at runtime[66]:

```
1  program hello_world_par2
2     implicit none
3     integer :: nThreads
4
5     write(*, '(a)', advance='no') "nThreads="
6     read*, nThreads
7
8     !$omp parallel num_threads(nThreads)
9        print*, "Hello, world of Modern Fortran! Parallel too!"
10    !$omp end parallel
11 end program hello_world_par2
```

Listing 5.22 $\boxed{\text{src/Chapter5/hello\_world\_par2.f90}}$

Note that we used the term *requested* above. As it turns out, for security reasons, the runtime *may allocate a lower number of threads than requested*. Therefore, if the number of threads appears within the code, one *should always check the actual number of threads*, using the $\boxed{\text{omp\_get\_num\_threads}}$-function, which we demonstrate later.

5.3.4.3 Assigning Work to Threads

The `parallel`-directive allows us to write multi-threaded OpenMP-programs. However, if we were to use only this, the programs we would obtain would not

---

[66] The $\boxed{\text{num\_threads}}$-clause overrides the environment variable $\boxed{\text{OMP\_NUM\_}}$ $\boxed{\text{THREADS}}$, if that is also set. Finally, note that the $\boxed{\text{omp\_set\_num\_threads}}$-subroutine can also be used to request a number of threads. For brevity, we do not use this in our examples.

run any faster! For example, in Listings 5.21 and 5.22, we were *not sharing the work* but rather *doing the same work, in parallel, multiple times*: the threads executed *exactly* the same instructions. This is, of course, not very useful—in practice, we *want each thread to execute a subtask*. In other words, we want each thread to execute a different code path or, for array-oriented problems, to apply the same instructions but to different sub-partitions of the array. In this section, we will discuss several methods for achieving this in OpenMP.

Differentiation by IDs and the SPMD pattern One method to assign different tasks to different threads with OpenMP is to manually divide the work, based on the thread ID. To illustrate, here is how we could extend the program from Listing 5.22, so that we get different messages from the master and worker threads:

```
1  program hello_world_par3
2    use omp_lib
3    implicit none
4    integer :: nThreads
5
6    write(*, '(a)', advance='no') "nThreads="
7    read*, nThreads
8
9    !$omp parallel num_threads(nThreads)
10     if( omp_get_thread_num() == 0 ) then
11       write(*, '(a,x,i0,x,a)') "Hello from MASTER (team has", &
12                 omp_get_num_threads(), "threads)"
13     else
14       write(*, '(a,x,i0)') "Hello from WORKER number", omp_get_thread_num()
15     end if
16   !$omp end parallel
17 end program hello_world_par3
```

Listing 5.23 | `src/Chapter5/hello_world_par3.f90`

First, in *line 2* we use the `omp_lib` module, which allows us to access the runtime library. For simplicity, we do not worry about sequential equivalence in this example. Due to the omp parallel directive, all threads will start by evaluating *line 10*, where the function `omp_get_thread_num` is used for the logical expression of the if-statement. The master thread will then execute the first write-statement (*lines 11–12*), where we use another OpenMP-function, `omp_get_num_threads`, to get the total number of threads in the team, including the master. As already noted, this number may well be different from nThreads, so we always have to check the actual number. Unlike the master thread, the workers will execute the other write-statement (*line 14*). This pattern for distributing work is also known as *single program, multiple data* SPMD, and it may look familiar to readers with some MPI[67] experience.

Since it assumes that the distribution of tasks is done by the programmer (based on the thread ID), the SPMD pattern is a quite general approach for parallel computing. The disadvantage, on the other hand, is that the code can become quite verbose, especially when the number of tasks is not exactly divisible by the number of threads. OpenMP includes many *worksharing constructs*, which greatly simplify this task.

Parallel `sections` A simple worksharing construct is sections, which is useful when we can identify a small (and fixed) number of subtasks in our algorithm.

---

[67] There, a similar idea is used for distributing the work, only that we refer to MPI ranks instead of thread IDs.

For example, assume that we have two tasks (A and B), and we want to run them in parallel (without being concerned with which thread executes which of the tasks). Our implementation based on `sections` would look like:

```fortran
subroutine doTaskA()
  implicit none
  write(*,'(a)') "Working hard on task A!"
end subroutine doTaskA

subroutine doTaskB()
  implicit none
  write(*,'(a)') "Working hard on task B!"
end subroutine doTaskB

program demo_par_sections
  implicit none

  !$omp parallel num_threads(2)
    !$omp sections
      !$omp section
        call doTaskA()
      !$omp section
        call doTaskB()
    !$omp end sections
  !$omp end parallel
end program demo_par_sections
```

Listing 5.24 | `src/Chapter5/demo_par_sections.f90`

Note that the `sections`-construct (*lines 15–20*) is embedded within a `parallel`-region (*lines 14–21*). Also, since we only have two tasks that could be executed in parallel, we use the clause $\boxed{\texttt{num\_threads(2)}}$ to limit the number of threads in the team. Finally, inside the `sections`-construct, the work for each $\boxed{\texttt{section}}$ (singular) is specified in a block of code preceded by the $\boxed{\texttt{!\$omp sections}}$-line. We present a more useful application of this construct in Sect. 5.3.5.2.

Parallel $\boxed{\texttt{do}}$ **loops** For many applications (especially in ESS), the bulk of the computing time is spent inside loops. Since OpenMP was initially designed with such applications in mind, it has extensive support for splitting iterations of loops between multiple threads, with the $\boxed{\texttt{do}}$ directive. As a simple first example, let us consider a program which applies an elemental function to a vector of values (this is also known as the *map pattern*—see, e.g., McCool et al. [19]):

```fortran
program demo_par_do_map
  implicit none
  real, dimension(:), allocatable :: arr
  real :: step = 0.1
  integer, parameter :: I4B = selected_int_kind(18)
  integer(I4B) :: numElems = 1E7, i

  arr = [ (i*step, i=1, numElems) ]

  !$omp parallel
    !$omp do
      do i=1, numElems
        arr(i) = sin( arr(i) )
      end do
    !$omp end do
  !$omp end parallel

  write(*,*) arr(numElems)
end program demo_par_do_map
```

Listing 5.25 | `src/Chapter5/demo_par_do_map.f90`

Similar to `sections`, the do-construct is embedded inside a `parallel`-region (*lines 10–16*). Immediately after the beginning `omp do` directive we must provide a loop to be parallelized across the threads in the current team. If we have nested loops, the construct will only affect the first loop. For this loop, the compiler will automatically create a private variable for each thread, to store its own loop index `i`. Also, each thread will receive a subset of the iteration space [1, *numElems*], without us having to do this manually (edge cases will also be handled by the compiler; for example, all the iterations will be executed, even if `numElems` is not exactly divisible by the number of threads in the team). For optimizing performance, OpenMP still allows the programmer to influence how the iteration space is sub-divided, using the `schedule`-clause. We do not elaborate here on this issue (pointers to additional resources can be found in Sect. 5.3.5.4).

Note that the closing directive (*line 15* above) could be skipped in principle, but we prefer to always specify it.

Applicability of `omp do`

For a loop to be correctly parallelizable using the `omp do` construct, most of the limitations we discussed previously for the `do concurrent` construct (Sect. 2.6.7.2) should also be satisfied. To summarize, there should be no inter-dependencies between the iterations of the loop, so that the compiler is free to execute those iterations in any order. Although (unlike for `do concurrent`) the compiler may not complain even if there is a violation of this principle, programmers need to be aware of this possible pitfall.

In Sect. 5.3.5.3, we will demonstrate how to use the `omp do` construct to parallelize the `LBM` solver we developed in the previous sections.

`single` Inside a `parallel`-region, it is possible to isolate some code so that it is executed only by a single thread. Although this may seem counter-intuitive,[68] it sometimes makes sense. A common usage is for initializing a global variable in which we store the actual number of threads in the current team,[69] as in:

```fortran
program demo_omp_single
   use omp_lib
   implicit none
   integer :: nThreads

   !$omp parallel
     !$omp single
       nThreads = omp_get_num_threads()
       write(*,'(2(a,x,i0,x),a)')"Thread", omp_get_thread_num(), &
         "says: team has", nThreads, "threads"
     !$omp end single

     ! remaining code within parallel-region executed by all
     write(*,'(a,x,i0,x,a)') "Thread", omp_get_thread_num(), &
```

---

[68] After all, we started the `parallel`-region to run in parallel!

[69] Note that calling the function `omp_get_num_threads` *outside* the `parallel`-region would simply return "1" (since only the master thread is active there).

```
15          "says: executing common code!"
16     !$omp end parallel
17 end program demo_omp_single
```

Listing 5.26 `src/Chapter5/demo_omp_single.f90`

Whichever thread (master or worker) "arrives" at the `single`-region first will execute the code inside the construct, while the other skip the code inside the construct and wait for that thread to finish. Afterwards, the entire team will execute the remaining code inside the `parallel`-region.

Compact forms for worksharing constructs A very common pattern when working with OpenMP is to have a `parallel`-region which simply wraps around a worksharing construct (such as `sections` or `do`). For such situations, OpenMP supports abbreviated notations (`parallel sections` and `parallel do`), to make the code more readable.

For example, here is how we could use this feature for the example in Listing 5.24:

```
!$omp parallel num_threads(2)
   !$omp sections
     !$omp section
       call doTaskA()
     !$omp section
       call doTaskB()
   !$omp end sections
!$omp end parallel
```

Listing 5.27 Verbose form of `omp sections` .

```
!$omp parallel sections num_threads(2)
   !$omp section
     call doTaskA()
   !$omp section
     call doTaskB()
!$omp end parallel sections
```

Listing 5.28 Equivalent, compact form of `omp sections` .

Similarly, the example in Listing 5.25 can also be made more concise:

```
!$omp parallel
   !$omp do
     do i=1, numElems
       arr(i) = sin( arr(i) )
     end do
   !$omp end do
!$omp end parallel
```

Listing 5.29 Verbose form of `omp do` .

```
!$omp parallel do
   do i=1, numElems
     arr(i) = sin( arr(i) )
   end do
!$omp end parallel do
```

Listing 5.30 Equivalent, compact form of `omp do` .

There is no compact version for omp single, since it makes no sense to start a team of threads and then to assign work only to a single thread from the team.

As a final note, while the compact versions are clearly easier to read, the reader should also remember that they are less general (when we need some code which is still *inside* the parallel-regions but *outside* the worksharing constructs, we have to use the verbose form).

5.3.4.4 Threads and Scope of Variables ("real Work" in OpenMP)

The reader may have noticed that, for the OpenMP-examples so far, we largely avoided working too much with variables inside parallel-regions.[70] Obviously, in real applications we need more flexibility, and OpenMP would not be very useful if it were so restrictive. However, before we present more realistic examples, we need to discuss some rules governing the scope of data. This knowledge will allow us to avoid conflict situations, such as having multiple threads trying to update the same variable at the same time (this scenario belongs to a class of problems unique to concurrent/parallel software, known as *race conditions*[71]). This is a crucial aspect, and it is also where OpenMP differs significantly from MPI.

Automatically shared **variables** Consider Fig. 5.5, where we illustrate some of the aspects related to data access in the OpenMP-model, along with sample code snippets. When a user launches a program, all code and data used by the program is grouped by the OS into a *process*. The threads of the process also reside within this context. From the point of view of the threads, unless specified otherwise, variables or constants are shared if they were declared:

1. prior to the parallel-region (but in the same program unit)
2. in the data section of an imported module (as a public entity)
3. with the save-attribute in a procedure which is called within the parallel-region.

For example, in Fig. 5.5, variable x would be shared by the threads, because it was declared in the same program unit as the parallel-region, and there is no clause to special "privatize" it.

Automatically private **variables** In addition to shared-data, the threads also have private-regions of memory,[72] which cannot be accessed by other threads

---

[70] In particular, we had *assignments* to variables only in Listing 5.25 (when each thread was guaranteed to write to different portions of the arr-array, at *line 13*), and in Listing 5.26 (where the assignment was performed by a single thread, since it occured within a omp single region, at *line 8*).

[71] In general, a *race condition* can occur whenever we have parallel tasks involving a variable, and at least one of the tasks is *writing* to that variable.

[72] This memory is usually allocated within a portion of memory known as the *stack*, where the local variables of a procedure would also reside (multiple threads are often supported by splitting the stack).

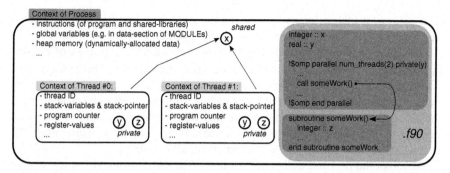

Fig. 5.5 Schematic of the OpenMP execution context (memory regions) for a program launched by the user

normally. In Fortran,[73] there are two common situations when variables automatically become part of this `private`-memory:

1. if they are local to procedures called from the `parallel`-regions, or
2. if the variable represents the index of a loop (do, *implied*-do, or `forall`) that is preceded by an omp do directive.

For example, the first rule applies to variable $\boxed{z}$ in Fig. 5.5: when threads will start executing the subroutine $\boxed{\text{someWork}}$, they will each get their own copy of $\boxed{z}$. On the other hand, if $\boxed{z}$ would have been declared with the $\boxed{\text{save}}$-attribute, it would be `shared` by all threads. The second rule applies to Listing 5.25, where the loop index needs to be `private` for the parallel function evaluation to work properly.

Also in the `private` memory areas, the OpenMP-runtime stores for each thread some internal bookkeeping information, such as the *thread ID* (which we already encountered) and an individual *program counter* (since different threads will generally execute different instructions in each cycle[74]).

Explicitly **controlling scope of variables** In addition to the *implicit* scoping rules mentioned above, OpenMP also allows programmers to further refine the data scope (i.e. to select what data is `shared` by the threads, and what is `private` to each thread). When none of the two cases from the previous paragraph applies, the implicit assumption in OpenMP is that variables are `shared`. This is often not the intended behavior, so programmers can also "privatize" variables, by adding them

---

[73] In C and C++, it is also allowed to have variable declarations inside the `parallel`-regions, and those are also made `private`. However, this mechanism is currently not supported in Fortran.

[74] Even if there are no divergent program flow paths (such as `if`s) inside the `parallel`-section, there is always some "system noise", so that threads are not guaranteed to work perfectly synchronized.

in a private-clause (or its variants firstprivate and/or lastprivate, described in Sect. 5.3.4.5), when the parallel-region is created.

We already had an example in Fig. 5.5, where we modified the scope of variable y with the clause private(y) , so that each thread gets a private copy of it.

As with automatically-private variables, the compiler will allocate for each thread a copy of the variable(s) listed inside the parentheses after the clause. Inside the list, even allocatable-arrays; in that case, if the array was allocated prior to the regions, it will also be allocated with the same shape inside the parallel-region (otherwise, it will remain un-allocated).

Although the private-directive is often essential, there are also a few related subtleties worth considering:

- Similar to how normal variables need to be initialized before using them in computations, *variables which were explicitly made* private *should be initialized before using them inside the* parallel-*region, unless the* firstprivate-*clause is used* (see Sect. 5.3.4.5).
- Also, the private-copies of variables exist only for the duration of the parallel-region – any information stored in the local variables will esentially vanish, unless it was either (a) saved into the larger context of the parent process (e.g. shared-variables), or (b) transferred back with the lastprivate-clause (see also Sect. 5.3.4.5).
- Finally, note that older implementations of OpenMP (prior to version 3.0) do not guarantee that, *after* the parallel-region completes, the original value of the variable that was "privatized" is still visible.

Another allowed scope clause is shared. As mentioned, this is the default behavior, so strictly speaking it is not necessary. However, sometimes it may be useful to specify this as a reminder for programmers, to avoid misinterpretations. Also, this is very useful when the default scope is changed, for debugging.

Instead of single variables, we can also pass lists of comma-separated variable names to the data clauses, to avoid repetition. For example:

```
integer : : x1, x2, x3
real : : y1, y2, y3

!$omp parallel private(x1, y1, x3) shared(x2, y2, y3)
  ...
!$omp end parallel
```

Note that the variables mentioned in the data clauses need to be declared prior to the parallel-*region, and in the same procedure.*

Default scope and debugging As already mentioned, the implicit access clause in OpenMP is shared. However, it is also possible to choose a *different default scope* (for a single directive at a time). This is done by adding a default(<policy_name>) to the directive, where for <policy_name> we have the following options:

- `shared` (redundant): all threads access the same memory location,
- `private`: create individual copies of variables for each thread,
- `firstprivate`: same as `private`, except that the individual copies are also *initialized, with the value prior to the region*)
- `none`: this will force us to explicitly set the scope for all variables inside the region, with the `private(<list_of_vars>)` and/or `shared(<list_of_vars>)` clauses mentioned before.

Using `default(private)` or `default(firstprivate)` may be useful when there are many variables to consider, and only few of them need to be `shared`. However, `default(none)` is particularly useful, especially while parallelizing a serial application or debugging, since it forces the programmer to think about how each variable is accessible by the threads (which is crucial for correctness). *We recommend to always start with* `default(none)`, *and then to explicitly specify the scope of each variable, until the code compiles. In a way, this facility is analogous to* `implicitnone`.

5.3.4.5 Information Transfer Into/out of a Thread's Context (`firstprivate` and `lastprivate`)

When a variable is made `private` to each thread for the duration of a `parallel`-region, the OpenMP-standard allows the new thread-local variables to have any random value – the assumption is that the programmer will take care of initializations (e.g. somewhere at the beginning of the region), before the values are used for any computations. Similarly, when the `parallel`-region ends, the values of any variables which are `private` to a thread are effectively lost.

Of course, for most real-world algorithms we *need* the threads to communicate with the larger context of the process. OpenMP accomodates this need with several mechanisms, of which we only demonstrate a few.

`firstprivate` **and** `lastprivate` We begin with two simple patterns which occur so frequently that OpenMP provides special support:

- **initializing variables with** `firstprivate`: It is often necessary to initialize a `private`-variable with the value of the variable before "privatization" (in the single-threaded region). This is the role of the `firstprivate`-clause, which is a superset of `private`.
- **propagating the "last" value with** `lastprivate`: A second common requirement is to propagate the "final" value of a `private`-variable outside a `parallel`-region, so that the next single-threaded region can use this value. This can be achieved with the `lastprivate`-clause. Note that this only works when there is a *natural sequential ordering of the tasks* (e.g. omp do and omp sections).

To illustrate these clauses (and how they differ from `private`), consider the following example:

```fortran
1   program demo_first_last_private
2      use omp_lib
3      implicit none
4      integer :: x=1, y=2, z=3, i=4
5
6      write(*,'(4(a,i0))')"A (serial) :: x =", x, &
7         ", y =", y, ", z =", z, ", i =", i
8
9      write(*,*) ! output-separator
10
11     !$omp parallel private(x) shared(y) &
12     !$omp firstprivate(z)
13        write(*,'(5(a,i0))') &
14           "B (parallel) :: Thread", omp_get_thread_num(), &
15           "says: x =", x, ", y =", y, ", z =", z, &
16           ", i =", i
17        ! assign to private variable
18        x = 2*omp_get_thread_num()
19
20        write(*,'(5(a,i0))') &
21           "C (parallel) :: Thread", omp_get_thread_num(), &
22           "says: x =", x, ", y =", y, ", z =", z, &
23           ", i =", i
24     !$omp end parallel
25
26     write(*,*) ! output-separator
27
28     !$omp parallel do shared(y)
29        do i=1, 42
30           y = y+i ! *** BUG! *** (data-race)
31        end do
32     !$omp end parallel do
33
34     write(*,'(4(a,i0))') "D (serial) :: x =", x, &
35        ", y =", y, ", z =", z, ", i =", i
36
37     !$omp parallel sections lastprivate(i)
38        !$omp section
39        i = 11
40        !$omp section
41        i = 22
42        !$omp section
43        i = 33
44     !$omp end parallel sections
45
46     write(*,'(4(a,i0))') "E (serial) :: x =", x, &
47        ", y =", y, ", z =", z, ", i =", i
48  end program demo_first_last_private
```

Listing 5.31 | src/Chapter5/demo_first_last_private.f90

On our test machine, this produces the following output:

```
1   $ OMP_NUM_THREADS=2 ./demo_first_last_private
2   A (serial) :: x = 1, y = 2, z = 3, i = 4
3
4   B (parallel) :: Thread 0 says: x = 1551671, y = 2, z = 3, i = 4
5   C (parallel) :: Thread 0 says: x = 0, y = 2, z = 3, i = 4
6   B (parallel) :: Thread 1 says: x = -1945985025, y = 2, z = 3, i = 4
7   C (parallel) :: Thread 1 says: x = 2, y = 2, z = 3, i = 4
8
9   D (serial) :: x = 1, y = 674, z = 3, i = 4
10  E (serial) :: x = 1, y = 674, z = 3, i = 33
```

Here is what happens to each of the variables (x , y , z , i) in Listing 5.31:

- x : At *line 11*, we declare this variable as `private`, which causes the compiler to instantiate a copy for each thread (we used 2 threads for brevity). However, there is no effort to initialize this variable, which is why random values are encountered at checkpoint B (*lines 13–16*). Then, each thread assigns a different value to this variable, as confirmed at checkpoint C (*lines 20–23*). Finally, after the first parallel-region ends (*line 24*), the original value[75] of the variable is shown (at checkpoints $\boxed{D}$ and $\boxed{E}$). Also, because $\boxed{x}$ is not explicitly mentioned in data clauses for the second and third parallel-regions, it becomes shared.

- $\boxed{y}$: For the first two parallel-regions (*lines 11 and 28*), we declare this variable as shared, which causes all threads to access the same memory location. Since the value is not updated in the first parallel-region, we get the initial value ($y=2$) at checkpoints $\boxed{B}$ and $\boxed{C}$. However, in the second parallel-region, at *line 30*, we update this shared value. This is a classic example of a *data race*, caused by the fact that the loop iterations are inter-dependent. While this do-loop would be perfectly valid when executed serially, the result becomes *undetermined* when running in parallel, because nothing stops here two threads from updating $\boxed{y}$ at the same time. Therefore, the result at checkpoint $\boxed{D}$ will be in general non-deterministic, and dependent on the number of threads (which we encourage the reader to try). The last parallel-region does not change $\boxed{y}$, so the same value is reported at checkpoint $\boxed{E}$. However, its status is also shared (but now due to implicit rules).

- $\boxed{z}$: The variable $\boxed{z}$ is declared as firstprivate on *line 12* which causes the initial value to be copied inside the private-versions of this variable, which the compiler creates for each thread. Note that this was *not* the case for $\boxed{x}$. For the rest of the program (*lines 26–47*), $\boxed{z}$ becomes shared due to the implicit rules.

- $\boxed{i}$: Finally, $\boxed{i}$ is silently shared for the first region. However, for the second region (*lines 27–32*), it becomes a private-variable, because it is the index of the loop. Since the loop has a pre-determined range of values to iterate through, there is no need for initialization. In the last region (*lines 37–44*), $\boxed{i}$ is declared as lastprivate, which will cause the value from the *sequentially last* task (33) to be copied outside, as reported at checkpoint $\boxed{E}$.

More complex communication As demonstrated in the previous example, there are several ways for allowing threads to communicate with the parent context. When applicable, firstprivate and lastprivate are excellent choices. Another common pattern is when threads need to cooperate and produce a unified value (e.g. a global sum, or extreme value). This is supported in OpenMP by the $\boxed{\text{reduction}}$-clause (see Sect. 5.3.5.4 for references).

Very often, however, the only viable choice is to use shared-variables. This is the case for many ESS applications, where each thread often operates on a sub-section of a large multi-dimensional array. OpenMP does not allow "privatizing" such subsections of arrays[76] (or of components of an ADT, for that matter). Therefore, it is up to the programmer to ensure that accesses to shared-data do not cause correctness problems (as it happened for variable $\boxed{y}$ in Listing 5.31). In general, such problems can be mitigated with synchronization constructs, of which

---

[75] Note that this is *not* guaranteed by OpenMP version 2.5 or lower.

[76] However, it is possible to define smaller arrays local to each thread, with threadprivate (see the references in Sect. 5.3.5.4 for details); this is more in the spirit of MPI, and may be used for porting OpenMP-programs to/from that technology [17].

there are many in OpenMP (`critical`, `atomic`, `barrier`, etc.). However, these techniques are beyond the scope of our text—in the following case studies, we will use `shared`-arrays, but we restrict the update operations, so that there are no conflicts or dependencies between the individual node updates for each timestep.

5.3.5 Case Studies for Parallelization

In this section, we provide more realistic use cases for OpenMP, by adding parallelism to the applications we presented in Sect. 4.1 (heat diffusion solver) and Sect. 4.3 (LBM-MRT solver for the *Rayliegh-Bénard* (RB) problem). Since both applications received further improvements earlier in this chapter, we choose those versions as starting points for parallelization.

5.3.5.1 Performance Optimization and Profiling

One of the important advantages of using OpenMP is that parallelism can often be added *incrementally*, by *profiling* the application after each significant change, to check where most of the computing time is spent.

A *profiler* is an application which can analyze our program, to characterize various aspects of its behaviour (performance "hotspots", call graph, etc.). One such tool is the *GNU Profiler* `gprof` (open source, available on most `Unix`-systems, but without dedicated support for OpenMP at the moment). Profiling a *serial* program with `gprof` involves repetitions of three basic steps:

1. **compile/link with `gprof` support**: On most `Unix` platforms, `gprof` requires us to add the `-pg` flag to *both* the compilation and linking stages. This will cause the final executable to contain additional code for tracking function call times.
2. **running the program**: For the second step, we need to run our program as usual. The main difference is that the program will also create `gmon.out`, which is a *binary file* (not human-readable) where the profiling result is stored.
3. **inspecting the result**: Last, we invoke the `gprof` program itself, which parses `gmon.out` and produces a human-readable summary. Several options are permitted at this stage, to display several aspects of the analysis. For our purposes here,[77] we will use the following syntax for this stage:

```
$ gprof -p -b ./program_name ./gmon.out
```

In addition to `gprof`, readers will probably find more advanced profiling tools, especially on HPC systems. Many such tools are supported by the hardware vendors

---

[77] We encourage the reader to check the official website for more information.

themselves, or by commercial software companies. For example, we use the *VTune Amplifier XE 2013* (`VTune`) for some of the later analyses at the end of this section.

Preserving program correctness When possible, it is very helpful to keep each optimization "iteration" *behavior-preserving* [18], i.e., to make sure that the program results are the same before and after adding more optimizations and/or parallelism.[78]

> **Caveat Emptor: no parallel I/O**
> Our parallelization case studies below only consider issues with parallelization of computations, and only write to disk the final state of the simulations. This causes the time spent on I/O to be insignificant, which simplifies our discussion here. Clearly, this assumption does not hold at all for transient model runs, for example. In such cases, it can become necessary to include I/O into the discussion of parallelization.

5.3.5.2 Example 1: Parallelizing the Heat Diffusion Application

As a first practical example, let us consider again the simple solver for the $2D$ heat diffusion equation, which we developed in Sect. 4.1 and extended in Sect. 5.2.1. In this section we describe how we can apply OpenMP to obtain a "low hanging fruit" improvement in performance.

Profiling of the *serial* program Before we invest any effort into parallelization, we need to determine unequivocally the hotspots in our program. Here, we do this using `gprof`. The three steps mentioned in Sect. 5.3.5.1 are applied here too, to the serial version of the application as in Sect. 5.2.1. First, we compile the program with profiling support:

```
$ gfortran -O2 -march=native -pg -o solve_heat_diffusion_v2{,.f90}
```

Then, we run the executable as usually:

```
$ ./solve_heat_diffusion_v2
```

This creates the file `gmon.out`, which we analyze with the command:

```
$ gprof -p -b ./solve_heat_diffusion_v2 ./gmon.out
```

---

[78] However, note that in typical ESS applications there *will* often be small fluctuations in the results, due to the non-associativity of floating-point operations [23]; therefore, some tolerances may need to be allowed when comparing results.

On our test system, we obtained the result:

```
Flat profile:

Each sample counts as 0.01 seconds.
  %   cumulative   self              self    total
 time   seconds   seconds   calls  us/call  us/call  name
51.52      5.48      5.48    9000   608.49   608.49  __solver_class_MOD_advancev
48.60     10.64      5.17    9000   574.00   574.00  __solver_class_MOD_advanceu
 0.00     10.64      0.00   90601     0.00     0.00  __solver_class_MOD_gettemp
 0.00     10.64      0.00       1     0.00     0.00  __config_class_MOD_createconfig
```

In the output, the first column displays the percentage of total time that was spent in each procedure (we can recognize some of the type-bound procedures of the Solver type in the last column). We notice that most of the effort is spent (in almost equal proportions) executing the subroutines advanceU and advanceV. Since these update the two sub-solution fields, which form the core of our algorithm, the profiling result will probably not surprise the reader. However, it is generally a good idea to profile often, since intuition often fails, especially in more complex applications.

Parallelization with OpenMP The reader may have already noticed that fields U and V can be updated at the same time, without any data conflicts. Therefore, the parallelization "effort" involves, in this case, nothing more than adding some directives for *parallel sections* in the subroutine run shown below (with indentation used to mark the nesting of OpenMP-constructs). Because the two tasks are already packaged as subroutines, the data scope for parallelization is quite simple—the "class" instance (this-variable) is shared by the threads, due to the default scoping rules:

```
191    subroutine run( this )  ! method for time-marching
192      class(Solver), intent(inout) :: this
193      integer(IK) :: k  ! dummy index (time-marching)
194
195      do k=1, this%mNumItersMax  ! MAIN loop
196        ! simple progress-monitor
197        if( mod(k-1, (this%mNumItersMax-1)/10) == 0 ) then
198          write(*, '(i5,a)') nint((k*100.0)/this%mNumItersMax), "%"
199        end if
200
201        ! NEW: OpenMP pragmas below
202        !$omp parallel num_threads(2)
203          !$omp sections
204          !$omp section
205            call this%advanceU()    ! task for 1st thread
206
207          !$omp section
208            call this%advanceV()    ! task for 2nd thread
209          !$omp end sections
210        !$omp end parallel
211        this%mCurrIter = this%mCurrIter + 1 ! tracking time step
212      end do
213    end subroutine run
```

Listing 5.32 | src/Chapter5/solve_heat_diffusion_v3/solve_heat_ diffusion_v3.f90 | (excerpt)

This modification alone brought a speedup of ~1.9 when using two threads[79] on our test machine, which is encouraging. However, it turns out to be more difficult to scale

---

[79] Of course, OpenMP-support also needs to be added at compile-time (e.g. by adding the −fopenmp flag for gfortran).

our chosen numerical algorithm beyond two threads, because the "semi-implicit" algorithm of Barakat and Clark [2] severely restricts the number of node-update sequences which lead to a correct result. An interesting class of such sequences form the basis of the *wavefront parallelization* technique [8], which could be used in this case. However, this is beyond the scope of this text.

The heat diffusion solver is a good case in point, showing that an algorithm which may perform well in serial can lead to difficulties during parallelization. For example, in this particular case, a parallel iterative algorithm (e.g. [9]) may lead to better utilization of the hardware.

5.3.5.3 Example 2: Parallelizing the LBM-MRT Application

For our last showcase for OpenMP, we will parallelize the LBM solver, which we introduced in Sect. 4.3 and extended in Sect. 5.2.2.4 (by adding netCDF-support). Unlike the previous example, this application can attain good scalability without having to restructure the entire algorithm (although, as we will show, there are still some potential traps along the way to good performance).

Profiling of the *serial* program Similar to the previous case study, the first step is to profile the serial version (lbm2d_mrt_rb_v2). As already noted, we only write output for the initial and final timesteps, since we focus here just on accelerating the raw computations; this is achieved by setting `numOutSlices=2` in file `src/Chapter5/lbm2d_mrt_rb_v2/lbm2d_mrt_rb_v2.f90`. To enable profiling, we need to append the `-pg` flag to variables FFLAGS and LDFLAGS (see file `src/Chapter5/lbm2d_mrt_rb_v2/Makefile.profiling`). We generate the human-readable version of the profile, using steps similar to the previous test case. A sample result on our system is[80]:

```
Flat profile:

Each sample counts as 0.01 seconds.
  %   cumulative   self              self
 time   seconds   seconds . us/call  us/call  name
 64.13    50.77     50.77 .  488.49   747.80  <advanceTimeMrtSolverBoussinesq2D>
 31.42    75.65     24.88 .    0.06     0.06  <calcLocalMomsMrtSolverBoussinesq2D>
  2.62    77.73      2.08 .    0.01     0.01  <calcLocalEqMomsMrtSolverBoussinesq2D>
  1.69    79.07      1.34 .   12.90    12.90  <getRawMacrosMrtSolverBoussinesq2D>
 . . . (more functions here, but which do not take much time) . . .
```

Most of the computational effort is spent on the procedure `advanceTimeMrt SolverBoussinesq2D`. This will be our primary target to parallelize since the next two procedures (`calcLocalMomsMrtSolverBoussinesq2D` and `calcLocalEqMomsMrtSolverBoussinesq2D`) are actually node-local computations (so there is not much to parallelize), and are in fact called only by this first procedure.

---

[80] To make the output fit in the page, we removed the column indicating the number of calls, and we also made the names more compact.

Parallelization with OpenMP Our simple approach for parallelizing this application will consist of parallelizing the spatial sweep in the `advanceTimeMrtSolver` `Boussinesq2D` type-bound procedure. The new version is shown below:

```
173    function advanceTimeMrtSolverBoussinesq2D( this ) result(res)
174      use omp_lib
175      class(MrtSolverBoussinesq2D), intent(inout) :: this
176      ! local vars
177      integer(IK) :: x, y, i, old, new, res
178      integer(IK), dimension(0:1) :: dest
179      real(RK) :: fluidMoms(0:8), tempMoms(0:4), &
180        fluidEqMoms(0:8), tempEqMoms(0:4)
181      integer, save :: numThreads = -9999
182
183      ! initializations
184      dest = 0; fluidMoms = 0._RK; tempMoms = 0._RK
185      fluidEqMoms = 0._RK; tempEqMoms = 0._RK
186      old = this%mOld; new = this%mNew
187
188      !$omp parallel &
189      !$omp shared(this, old, new, numThreads) private(x, y, i) &
190      !$omp firstprivate(fluidMoms, tempMoms, fluidEqMoms, tempEqMoms, dest)
191
192      !$omp single
193      if( numThreads == -9999 ) then
194        numThreads = omp_get_num_threads()
195      end if
196      !$omp end single
197
198      !$omp do
199      do y=1, this%mNy
200        do x=1, this%mNx
201          call this%calcLocalMomsMrtSolverBoussinesq2D(x, y, fluidMoms, tempMoms)
202
203          ! add 1st-half of force term (Strang splitting)
204          fluidMoms(2) = fluidMoms(2) + this%mAlphaG*0.5_RK*tempMoms(0)
205
206          ! save moments related to output
207          this%mRawMacros(x, y, :) = &
208            [ fluidMoms(0), fluidMoms(1), fluidMoms(2), tempMoms(0) ]
209
210          call this%calcLocalEqMomsMrtSolverBoussinesq2D( dRho=fluidMoms(0), &
211            uX=fluidMoms(1), uY=fluidMoms(2), temp=tempMoms(0), &
212            fluidEqMoms=fluidEqMoms, tempEqMoms=tempEqMoms )
213
214          ! collision (in moment-space)
215          fluidMoms = fluidMoms - this%mRelaxVecFluid * (fluidMoms - fluidEqMoms)
216          tempMoms  = tempMoms  - this%mRelaxVecTemp * (tempMoms - tempEqMoms)
217
218          ! add 2nd-half of force term (Strang splitting)
219          fluidMoms(2) = fluidMoms(2) + this%mAlphaG*0.5_RK*tempMoms(0)
220
221          ! map moments back onto DFs...
222          ! ...fluid
223          do i=0, 8
224            this%mDFs(i, x, y, old) = dot_product( M_INV_FLUID(:, i), fluidMoms )
225          end do
226          ! ...temp
227          do i=0, 4
228            this%mDFs(i+9, x, y, old) = dot_product( N_INV_TEMP(:, i), tempMoms )
229          end do
230
231          ! stream to new array...
232          ! ...fluid
233          do i=0, 8
234            dest(0) = mod(x+EV_FLUID(1, i)+this%mNx-1, this%mNx)+1
235            dest(1) = y+EV_FLUID(2, i)
236            ! STREAM (also storing runaway DFs in Y-buffer space)
237            this%mDFs(i, dest(0), dest(1), new) = this%mDFs(i, x, y, old)
238            if( dest(1) == 0 ) then
239              if( EV_FLUID(2, i) /= 0 ) then
240                ! apply bounce-back @bottom
241                this%mDFs(OPPOSITE_FLUID(i), x, y, new) = &
242                  this%mDFs(i, dest(0), dest(1), old)
243              end if
244            elseif( dest(1) == this%mNy+1 ) then
245              if( EV_FLUID(2, i) /= 0 ) then
246                ! apply bounce-back @top
247                this%mDFs(OPPOSITE_FLUID(i), x, y, new) = &
248                  this%mDFs(i, dest(0), dest(1), old)
249              end if
250            end if
251          end do
252          ! ...temp
253          do i=0, 4
254            dest(0) = mod(x+EV_TEMP(1, i)+this%mNx-1, this%mNx)+1
255            dest(1) = y+EV_TEMP(2, i)
256            ! STREAM (also storing runaway DFs in Y-buffer space)
257            this%mDFs(i+9, dest(0), dest(1), new) = this%mDFs(i+9, x, y, old)
258            if( dest(1) == 0 ) then
259              ! apply anti-bounce-back @bottom
260              this%mDFs(OPPOSITE_TEMP(i)+9, x, y, new) = &
261                -this%mDFs(i+9, dest(0), dest(1), old) + &
262                2._RK*sqrt(3._RK)*this%mDiffusivity*this%mTempHotWall
263            elseif( dest(1) == this%mNy+1 ) then
264              ! apply anti-bounce-back @top
265              this%mDFs(OPPOSITE_TEMP(i)+9, x, y, new) = &
266                -this%mDFs(i+9, dest(0), dest(1), old) + &
267                2._RK*sqrt(3._RK)*this%mDiffusivity*this%mTempColdWall
```

```
268                        end if
269                    end do
270                end do
271            end do
272            !$omp end do
273            !$omp end parallel
274
275            ! swap 'pointers' (for lattice-alternation)
276            call swap( this%mOld, this%mNew )
277
278            res = numThreads
279        end function advanceTimeMrtSolverBoussinesq2D
```

Listing 5.33 | `src/Chapter5/lbm2d_mrt_rb_v3/MrtSolverBoussines`

`q2D_class.f90` (excerpt)

A `parallel`-region is started at *lines 188–190*, and closed at *line 273*, to surround
the spatial sweep. Unlike the previous case study, here we have more variables for
which we need to clarify the scope. At *lines 189–190*, we declare as `private` or
`firstprivate` the variables used by each thread for storing intermediate results.
As `shared`, we have the solver instance (`this`), the variables which keep track
of the `old` and `new` lattices, as well as a variable for storing the actual number of
threads (`numThreads`, discussed later).

Within the `parallel`-region, we have a loop construct (starting at *line 198* and
ending at *line 272*), which causes the different threads to work on (non-overlapping)
sub-ranges of the *y*-direction. In principle, the loop construct could also have
been added around the inner *x*-loop. However, since each worksharing construct
in OpenMP introduces some overhead, it is usually a good idea to parallelize the
outermost loop possible, so that the workload for each thread is large enough to
mask that overhead.

For studying the performance as a function of the number of threads, our program
needs to inquire the OpenMP runtime system, to find out the *actual* number of
threads in the team. This number is stored inside the `numThreads`-variable, which
is declared at *line 181*, with the `save`-attribute. In *lines 193–195*, we have some code
which will update the variable during the first iteration. Since this is a `shared`-
variable, we surround the update with a `single`-construct. Finally, note that we
transformed `advanceTimeMrtSolverBoussinesq2D` into a `function`, so
that we can return the number of threads to the parent `RBenardSimulation`.
Because we use `omp_get_num_threads` at *line 194*, we also need to use the
`omp_lib`-module (*line 174*).

The `RBenardSimulation` "class" also needs to be modified slightly, to make
the performance-reporting code aware of OpenMP. Specifically, we change the
`runRBenardSimulation` method to:

```
 89    subroutine runRBenardSimulation( this )
 90        use omp_lib
 91        class(RBenardSimulation), intent(inout) :: this
 92        integer(IK) :: currIterNum, realNumThreads
 93        real(RK) :: tic, toc, numMLUPS ! for performance-reporting
 94
 95        tic = omp_get_wtime() ! parallel
 96
 97        ! MAIN loop (time-iteration)
 98        do currIterNum=1, this%mNumItersMax
 99            ! simple progress-monitor
100            if( mod(currIterNum-1, (this%mnumitersmax-1)/10) == 0 ) then
101                write(*, '(i5,a)') nint((currIterNum*100._RK)/this%mnumitersmax), "%"
102            end if
103
104            realNumThreads = this%mSolver%advanceTime()
```

```
105        call this%mOutSink%writeOutput( this%mSolver%getRawMacros(), currIterNum )
106     end do
107
108
109    toc = omp_get_wtime() ! parallel
110
111    numMLUPS = this%mNumItersMax*real(this%mNx*this%mNy, RK) / (1.0e6*(toc-tic))
112    write(*,'(/,a,f0.2,a)') "Performance Information: achieved", &
113        numMLUPS, "MLUPS (mega-lattice-updates-per-second)"
114    write(*,'(a,i0,a,f0.4,a)') "[ <nThreads>", realNumThreads, &
115        "</nThreads> <perf>", numMLUPS, "</perf> ]"
116 end subroutine runRBenardSimulation
```

Listing 5.34 | `src/Chapter5/lbm2d_mrt_rb_v3/RBenardSimulation_` `class.f90` (excerpt)

At *lines 95* and *109*, we use `omp_get_wtime` (instead of the subroutine `cpu_time`), to measure correctly the elapsed "wall clock" time for our application—without this modification, we would get instead the sum of the individual times for each thread, which would probably be comparable to the serial execution time. Again, since `omp_get_wtime` is a library procedure, we use `omp_lib` (*line 90*). The other small changes (also related to performance reporting) are the introduction of the `realNumThreads` and `numMLUPS` variables. Finally, at *lines 114–115*, we write the actual number of threads and the performance (in millions of lattice-node updates per second), using a special syntax. This is necessary for the `Python` script `omp_scaling_test.py` (also in the code repository), which is used for testing parallel scaling of performance, for a given range of (target) number of threads, and by repeating each simulation for a (configurable) number of times, to reduce measurement noise.

With these modifications, the performance of the code on our test system scales as follows[81]:

```
$ ./omp_scaling_test.py ./build/lbm2d_mrt_rb_v3 10 1 4
1 5.63824
2 10.75429
3 15.24982
4 19.12396
```

Listing 5.35 Scaling test for Version 3 of our LBM-MRT application. For reducing fluctuations in the performance figures, we use the `omp_scaling_test.py` script, which runs the simulation 10 times, for values of OMP_NUM_THREADS between 1 and 4. In the command output, the first column represents actual numbers of threads, and the second one the performance of the application (in millions of lattice-node updates per second).

Exercise 22 (*Querying for the number of threads*) For our second case study (see Listings 5.33 and 5.34) we made some effort to query the runtime system, and then propagate the number of threads to higher levels of our application.

---

[81] Note that we only wrote output for the initial and final timesteps; output writing will generally degrade scalability, due to Amdahl's law.

Explain why this is necessary; why not just call `omp_get_num_threads` from the `runRBenardSimulation`-procedure? Experiment to see what number of threads is reported if we did this instead.

5.3.5.4 More OpenMP to Explore

This concludes our brief coverage of OpenMP. Naturally, we only presented a small subset of the features available to the user (even if we only considered Version 3.1). From the (long) list of features which we *did not* explain but which may be crucial for many applications, we can mention:

- support for *dynamic parallelism* (`task`-construct),
- reductions,
- *explicit synchronization* (`barrier`, `critical`, etc.),
- "privatization" of global data (`threadprivate`), or
- techniques for performance optimization (load balancing, memory model, affinity, etc.).

The interested reader can consult, for example, Hager and Wellein [8] (an introduction to parallelization in general), Chapman et al. [6] and Chandra et al. [5] (for more on OpenMP), or Mattson et al. [18] and McCool et al. [19] (for related software engineering issues).

Also, at the time of writing, Version 4.0 was already published, which offers many more features worth considering, especially as compiler support matures.

5.4 Interoperability with C

Although many applications are written in a single language, a fact is that various programs and libraries were written by different programmers, with different preferences for specific languages. Besides subjective reasons such as individual expertise and preferences of the programmer, this variety also reflects the fact that different programming languages have their own strengths and weaknesses. For example, compiled languages like Fortran and C/C++ are suitable for performance-critical code, while an interpreted language (like R, Python, or MATLAB) would be preferred, e.g., for interactive data analysis.[82] For the application developer, there are often good reasons to *combine* different languages. Since the C programming language

---

[82] Due to the need to interpret code at runtime, scripting languages often introduce some performance penalty. However, in many situations (data analysis, algorithm prototyping, etc.) it is perfectly acceptable to trade some performance for higher programmer productivity. Also, many scripting languages allow some form of code compilation, so the distinction is not so clear-cut.

is so popular for system-level programming, many of the other languages include support for C; the C-layer can also be used to "bridge" two non-C languages.

In this section, we will briefly discuss about how Fortran can cooperate with C. In particular, we will discuss the situation where the program is written mainly in one language, but we also need to call a procedure written in the other language.[83]

Prior to the 2003 version of the language standard Fortran did not have any official support for interoperability with C. Despite this fact, however, some developers still managed to create applications that mixed code from the two languages, using vendor- or OS-specific extensions, external to the language standard. While such solutions work, they were not very portable.

To fix the portability problems, Fortran 2003 introduced a standard, cross-platform mechanism for interoperability. First, the new standard requires any compliant compiler to define a *companion C-compiler*. Since many vendors (as well as the gcc[84]) provide compilers in "tuples" (for Fortran, C and C++), there is usually a natural choice for the companion compiler.

The role of the new Fortran extensions is to instruct the Fortran compiler to generate (or to accept) code which agrees with the low-level conventions of the companion C compiler. Below, we demonstrate via examples some basic usage of these new facilities.

Vendor-specific build instructions
While the *code* for the next examples is portable, the instructions to compile and link the code are not—we demonstrate with gfortran and gcc. When using a different compiler-suite, the instructions should be adapted accordingly.

5.4.1 Crossing the Language Barrier with Procedures Calls

As a first example of interoperability, we consider how a program written in one language can invoke a procedure written in the other language. We present both cases (i.e. Fortran calls C and C calls Fortran). To keep this first version simple, our main programs will simply invoke the procedure, and the procedure will display some

---

[83] It is also possible (and common) to combine *complete programs* written in different languages, by exchanging information via files on disk or interprocess communication mechanisms (such as Unix *pipes*), and steering the execution of the programs through some scripts (e.g. shell scripts). However, here we refer to the case when we want to link together object files obtained from different compiled languages.

[84] For many compiler suites (including gcc), the Fortran and C compilers actually have common components, with different programming languages being supported by different frontends, which translate the code into a language-neutral intermediate representation.

message to standard output; later, we will iterate on these examples, to demonstrate more functionality.

5.4.1.1 Fortran Calls C (V1)

First, consider the situation when the main-program is written in Fortran, and the function we want to call is implemented in C, as follows:

```
5   #include <stdio.h>
6
7   void test_proc_c_v1() {
8     printf("Hello from \"test_proc_c_v1(C)\","
9            "invoked from \"demo_fort_v1(Fort)\"!\n");
10  }
```

Listing 5.36 src/Chapter5/interop/f_calls_c_v1/test_proc_c_
v1.c

There is nothing special about this function-definition—it takes no arguments, and simply displays a message to stdout. Using gcc under Linux, we can compile (*without* linking) the code with:

```
$ gcc -c test_proc_c_v1.c
```

The corresponding main-program (written in Fortran) is shown below:

```
5   program demo_fort_v1
6     implicit none
7     ! IFACE to C-function.
8     interface
9       subroutine test_proc_c_v1() &
10         bind(C, name='test_proc_c_v1')
11      end subroutine test_proc_c_v1
12    end interface
13
14    call test_proc_c_v1()    ! Fort -call-> C
15  end program demo_fort_v1
```

Listing 5.37 src/Chapter5/interop/f_calls_c_v1/demo_fort_v1.
f90

This is a little more interesting, since we have some new elements. In order to allow the Fortran compiler to perform error checking, we explicitly define the interface of the C function, with an interface-block (*lines 8–12*). This is similar to what we discussed in Sect. 3.2.3, with the addition of the bind-attribute (*line 10*), which causes the Fortran compiler to produce object code which is compatible with the conventions of the companion C compiler. Inside the parentheses of the bind-clause, the first element should be C . In principle, the second element (corresponding to name) is optional. However, we prefer to always specify it, even if the string is identical to the name of the procedure within the interface-block, as in Listing 5.37. This avoids potential problems due to mixed letter-case.[85]

---

[85] Remember that case variations are generally discarded by Fortran compilers, but not so by C compilers. Therefore, the argument corresponding to the name-keyword *is case-sensitive!*

We can compile (again, *without* linking) the code with:

```
$ gfortran -c demo_fort_v1.f90
```

Linking the final executable The two compilation commands above would have produced two object files. As discussed in Sect. 5.1, we can invoke the linker, to combine the object files (and the external libraries they need) into an executable. On our platform, the command is:

```
$ gfortran -o demo_fort_v1 demo_fort_v1.o test_proc_c_v1.o -lc
```

Note that we also link against the C standard library ($\boxed{\texttt{libc}}$ on Linux), which is necessary for `printf`.[86]

5.4.1.2 C Calls Fortran (V1)

Next, consider the reverse situation, when the main-program is written in C, but we need to invoke a `subroutine` written in Fortran—for example:

```
5  subroutine test_proc_fort_v1() &
6         bind(C, name='test_proc_fort_v1')
7     write(*,'(a)') 'Hello from "test_proc_fort_v1(Fort)",&
8            &invoked from demo_c_v1(C)"!'
9  end subroutine test_proc_fort_v1
```

Listing 5.38 $\boxed{\texttt{src/Chapter5/interop/c\_calls\_f\_v1/test\_proc\_}}$ $\boxed{\texttt{fort\_v1.f90}}$

To make the procedure callable from C-code, we need to specify again the bind-attribute (*line 6*). We generate the corresponding object file with:

```
$ gfortran -c test_proc_fort_v1.f90
```

The corresponding main-program (written in C) is:

```
5  #include <stdlib.h>
6
7  /* declaration of Fortran procedure */
8  void test_proc_fort_v1(void);
9
10  int main() {
11      test_proc_fort_v1();   /* C -calls-> Fort */
12
13      return EXIT_SUCCESS;
14  }
```

Listing 5.39 $\boxed{\texttt{src/Chapter5/interop/c\_calls\_f\_v1/demo\_c\_v1.c}}$

---

[86] This library is also automatically added by the compiler, so $\boxed{\texttt{-lc}}$ may be skipped in this case. However, we added it explicitly here, to facilitate comparison with the next example (Sect. 5.4.1.2).

Since the procedure is external to the translation unit, we need a forward declaration for it (*line 8*), just like we needed to add an `interface`-block in Listing 5.37. As usual, we generate the object file with:

```
$ gcc -c demo_c_v1.c
```

Linking the final executable Similar to Sect. 5.4.1.1, we invoke the linker (now through the C compiler, to obtain the final executable:

```
$ gcc -o demo_c_v1 demo_c_v1.o test_proc_fort_v1.o -lgfortran
```

Note that we now need to link in the Fortran standard library (`libgfortran`), for our invocation of `write` to work (Listing 5.38, *line 7*).

5.4.2 Passing Arguments Across the Language Barrier

For our first examples of interoperability, we demonstrated how to invoke procedures from the other language, without any arguments. However, in most interesting situations we *do* need to pass data between the two languages, so we discuss this here. For brevity, we only demonstrate two types of variables—a simple `integer`-scalar, and a 2*D* fixed-size array of `real`-values.

5.4.2.1 Interoperable Data Types

Any variables that are shared in some way with C obviously need to be of types which are accepted by the C compiler. From the Fortran side, we ensure this is the case by selecting special `kind`s for the intrinsic types. These special `kind` type parameters are defined within the intrinsic module `iso_c_binding`. With the exception of unsigned `integer`-types (which are not supported in Fortran), we have there `kind`-values for translating most common types; for example, `integer(c_int)` in Fortran is compatible with `int` in C, `integer(c_long)` with `long`, `real(c_float)` with `float`, `real(c_double)` with `double`, etc. Note that the compiler may not support interoperability for all of the types; in such situations, a negative `kind`-value will be returned, which will cause a compilation error when used later for variable declarations (see, e.g., Metcalf et al. [21] for a discussion of possible negative values and their meanings).

5.4.2.2 Fortran Calls C (V2)

We now extend the example from Sect. 5.4.1.1, adding a scalar and an array as procedure arguments. The new version of the C function is:

```c
#include <stdio.h>

void test_proc_c_v2(int n, double arr[3][2]) {
    int i, j;

    printf("Hello from \"test_proc_c_v2(C)\","
           "invoked from \"demo_fort_v2(Fort)\"!\n");
    printf("n = %d\n", n);
    for(j=0; j<3; j++) {
        for(i=0; i<2; i++) {
            printf("arr[%d,%d,%d] = %8.2f\n", j, i, arr[j][i]);
        }
    }
}
```

Listing 5.40 src/Chapter5/interop/f_calls_c_v2/test_proc_c_v2.c

At *line 12* we print the received value for n, and the nested loops at *lines 13–17* do the same for the array arr.

The corresponding main-program (Fortran) is:

```fortran
program demo_fort_v2
    use iso_c_binding, only: c_int, c_double
    implicit none

    integer :: i, j ! dummy indices
    integer(c_int) :: n_fort = 17
    real(c_double), dimension(2,3) :: arr_fort

    ! IFACE to C-function.
    interface
        subroutine test_proc_c_v2(n_c, arr_c) &
            bind(C, name='test_proc_c_v2')
            use iso_c_binding, only: c_int, c_double
            integer(c_int), intent(in), value :: n_c
            real(c_double), dimension(2,3), intent(in) :: arr_c
        end subroutine test_proc_c_v2
    end interface

    ! initialize 'arr_fort' with some data
    do j=1,3
        do i=1,2
            arr_fort(i,j) = real(i*j, c_double)
        end do
    end do

    call test_proc_c_v2(n_fort, arr_fort)  ! Fort -call-> C
end program demo_fort_v2
```

Listing 5.41 src/Chapter5/interop/f_calls_c_v2/demo_fort_v2.f90

Obviously, the data to be passed to the procedure needs to be declared, with C-compatible kinds, and initialized (*lines 10–11* and *23–28*).

There are some differences related to array support in Fortran versus C, which become relevant here:

- Because array storage order is different in Fortran and C ("column-major" vs "row-major"), the list of sizes along each dimension needs to be reversed for multi-dimensional arrays (compare *lines 11* and *19* in Listing 5.41 with *line 7* in Listing 5.40).

- Also, whereas Fortran allows arbitrary lower and upper bounds for the array indices[87] (with the lower bound defaulting to $\boxed{1}$ and upper bound to the size along that dimension), in C array indices have $\boxed{0}$ as lower bound and $\boxed{\texttt{sizeOf}}$ $\boxed{\texttt{Dimension}-1}$ as upper bound.

At *lines 17–19* in Listing 5.41, we update the `interface`-block, to account for the new arguments. The declarations also need to use the C-compatible `kind`-parameters. Also, it is important to notice that, *whereas procedure arguments are by default passed-by-reference in Fortran, the default passing mechanism in C is by-value.* This is why we need the additional `value`-type attribute (*line 18*)—otherwise, the C function would receive a pointer instead of the actual integer value.

5.4.2.3 C Calls Fortran (V2)

For the reverse scenario, when the main-program is written in C, we can invoke a Fortran `subroutine` such as:

```fortran
 5   subroutine test_proc_fort_v2(n, arr) &
 6         bind(C, name='test_proc_fort_v2')
 7     use iso_c_binding, only: c_int, c_double
 8     integer(c_int), intent(in), value :: n
 9     real(c_double), dimension(2,3), intent(in) :: arr
10     integer :: i, j ! dummy indices
11
12     write(*,'(a)')'Hello from"test_proc_fort_v2(Fort)",&
13         &invoked from demo_c_v2(C)"!'
14     write(*,'(a,i0)') "n =", n
15     do j=1,3
16       do i=1,2
17         write(*,'(2(a,i0),a,f8.2)') "arr[", i, ",", j, "] =", arr(i,j)
18       end do
19     end do
20   end subroutine test_proc_fort_v2
```

Listing 5.42 `src/Chapter5/interop/c_calls_f_v2/test_proc_` `fort_v2.f90`

Most observations from the previous section also apply here, including the `value` type attribute, which we now have to specify inside the `subroutine` (*line 8*).

The corresponding C main-program, and the corresponding procedure forward declaration, are:

```c
 5   #include <stdlib.h>
 6
 7   /* declaration of Fortran procedure */
 8   void test_proc_fort_v2(int n_f, double arr_f[3][2]);
 9
10   int main() {
11     int i, j, n_c=17;
12     double arr_c[3][2];
13
14     /* initialize 'arr_c' with some data */
15     for(j=0; j<3; j++) {
16       for(i=0; i<2; i++) {
```

---

[87] As long as they are representable `integers` and the resulting array fits into memory.

```
17          arr_c[j][i] = (double) (i+1)*(j+1);
18      }
19    }
20
21    test_proc_fort_v2(n_c, arr_c);   /* C -calls-> Fort */
22
23    return EXIT_SUCCESS;
24 }
```

Listing 5.43 | `src/Chapter5/interop/c_calls_f_v2/demo_c_v2.c`

This concludes our introduction to interoperability issues. In practice, the reader may also encounter more advanced scenarios, which we do not cover here. For example, it is also possible to pass between Fortran and C character-strings and dynamic arrays, or to make global data interoperable. Some additional type definitions and intrinsic procedures (also defined in the iso_c_binding-module) are relevant for such tasks—for more information, we refer to other texts (such as Clerman and Spector [7], Markus [15], or Metcalf et al. [21]).

5.5 Interacting with the Operating System (OS)

For a long time, Fortran programs had no standard mechanism for interacting with the OS, although vendor-specific mechanisms were available. To remove this potential source of portability problems, recent versions of the standard added some intrinsic procedures to streamline this process. In this section, we use some of these procedures, to demonstrate passing command line arguments and launching ("forking") another program directly from a running Fortran application.

5.5.1 Reading Command Line Arguments (Fortran 2003)

It is often useful to provide some additional information when executing a program. Considering, for instance, the simple solver for the heat diffusion equation (Sect. 4.1), it would be more convenient for the user to be able to specify the grid resolution parameter when launching our program. Also, a common practice is for programs to report a summary about the allowed CLI-arguments, when invoked with the single argument $\boxed{--\texttt{help}}$ (Unix[88]), as in:

```
$ ./program_name --help
```

There are two possible approaches for obtaining the command line arguments from the Fortran runtime system.

Option (A): read entire invocation command line As a first option, it is possible to obtain from the runtime system the whole command line, including the name of

---

[88] In Windows, a slash ($\boxed{/}$) is often used instead of the dash ($\boxed{-}$).

the program and the complete list of arguments. This information can be obtained by calling the intrinsic subroutine get_command, which has the syntax:

```
call get_command([command=]string\_val, [length=]string\_len, &
     [status=]cmd_stat )
```

where the arguments (all of them optional) represent:

- command=*string_val* : here, *string_val* is a string (intent(out)), which will be set to the value of the complete command line; depending on whether the variable is longer or shorter than the actual command line, zero-padding or truncation will be applied (so ensure a sufficient length is reserved for this string)
- length=*string_len* : is an integer (intent(out)), which will be set to the length of the complete command line
- status=*cmd_stat* : is another integer, used for detecting various error conditions (set to 0 if no error occurred, to −1 if the command did not fit in the string *string_val*, or to > 0 if the command could not be retrieved due to another problem)

Since this subroutine provides the "raw" command line, it has the drawback of forcing programmers to write their own code for parsing the command string.

Option (B): read arguments one-by-one The second method for obtaining the command line arguments consists of first asking the number of arguments, by calling the intrinsic function command_argument_count. This returns an integer, with the number of CLI-arguments, *not including the program name*. With this information, the programmer can then retrieve individual arguments based on an index, with the get_command_argument-subroutine, with the syntax:

```
call get_command_argument( [number=]arg_idx, [value=]string\_val, &
     [length=]string\_len, [status=]cmd_stat )
```

where the arguments (all optional, except the first one) represent:

- number=*arg_idx* : integer (intent(in)), containing the index of the argument to be retrieved (value of 0 can also be used, to get the name of the program itself)
- value=*string_val* : string (intent(out)), where the value of the argument will be placed (again, truncated or zero-padded if the argument is larger or shorter than the string's length)
- length=*string_len* and status=*cmd_stat* have similar roles as in the first method (but now applied to individual CLI-arguments)

With this second approach, the task of splitting the list of arguments (also known as *tokenization* in the programming jargon) is accomplished by the compiler. However, the programmer still has to write some code to validate and interpret the arguments (which can be somewhat tedious for options which accept values, e.g., something

like $-n=123$).[89] We demonstrate the two methods of reading arguments in the file
`src/Chapter5/reading_cli_arguments.f90` (see the source code repository).

5.5.2 Launching Another Program (Fortran 2008)

Another aspect of OS-interaction is the ability of a Fortran program of asking the OS to execute another program. This can be used in creative ways, such as for interacting with the user via GUI-dialogs (for example, using `zenity` in conjunction with shell scripts), or for providing immediate visualizations of the program results. We describe the latter scenario here.

The relevant intrinsic procedure is `execute_command_line`, with the general calling syntax:

```
call execute_command_line([command=]cmd, [wait=]wait_flag, &
        [exitstat=]e_stat, [cmdstat=]c_stat, [cmdmsg=]c_msg )
```

where we used square brackets to indicate optional keywords that can be used. The only mandatory argument is *cmd*, which needs to be a character string (`intent(in)`), containing the command to be transferred to the OS (such as "ls" for displaying a listing of the present directory in `Unix` – "dir" in Windows). The other arguments are optional, but are necessary for error checking, or for executing the command contained in *cmd* asynchronously. We only describe the ones relevant for the former (so the *wait_flag* can be omitted[90]):

- `exitstat=e_stat` : here, *e_stat* is an `integer` (`intent(inout)`), which will be set to the launched program's exit status (usually zero representing success, and non-zero – failure); this variable is useful for checking error codes issued by the launched program itself
- `cmdstat=c_stat` : *c_stat* is another `integer` (`intent(out)`), this time generated by the Fortran runtime system, which returns 0 if `execute_command_line` itself executed with no errors, -1 is this feature is not supported, or a value > 0 if another error occured
- `cmdmsg=c_msg` : *c_msg* is a character string (`intent(inout)`), which should contain more information related to the error reported in *c_stat* (if any)

The optional arguments are provided, as usual, to allow the programmer to recover from exceptional situations (if they are absent, the program will simply be aborted). To demonstrate the use of this feature, we provide an example (file `src/Chapter5/`

---

[89] C++ programmers do not need to do this, since they can use libraries such as `Boost.Program_Options`.

[90] Although the ability to launch programs asynchronously looks very appealing (indeed, this can even be viewed as a primitive form of parallelization), its usefulness is limited in practice, since there is currently no standard mechanism to check if the program actually terminated and, if so, what exit status was returned. See [21] for more details on this feature.

`test_launching_external_programs.f90` in the source code repository).

5.6 Useful Tools for Scaling Software Projects

In closing, we briefly mention several other tools that may become useful for your projects. Note that, depending on the current (expected) size of the software, not all technologies mentioned here may pay off.

5.6.1 Scripting Languages

In this book we focus on using Fortran as a tool for solving computational problems. However, it is not feasible to write all types of programs in Fortran. The language is not suitable, for example, for applications where code is changing rapidly, and there is a need to quickly test the outcome. Likewise, it is more economic to choose another language when extensive functionality from a specific problem domain (such as graphics or process manipulation) is necessary, which is not available in Fortran, or in a 3rd-party library callable from Fortran.

A common practice to resolve this tension is to develop multi-language applications, so that the strengths of each language can be exploited. We already discussed some form of this in Sect. 5.4. Given the supremacy of C and C++ in high-performance systems programming, there is a large set of useful libraries which become available to Fortran programmers through such an inter-language bridge. However, this does not solve the problem of applications where requirements change rapidly (such as exploratory visualization and data analysis), and can also increase the complexity of the final applications.

Another common combination in relation to Fortran is to use a scripted language for the tasks which are more cumbersome to implement directly in Fortran (such as file system manipulation, or computational steering); the scripts then delegate the numerically intensive tasks to Fortran programs. Most of the scripted languages do not need to be compiled, which allows to immediately get feedback from individual commands. The traditional scripting languages in Unix are *shell scripts* (like bash, zsh, ksh, or tcsh). For ESS applications, which need to run on supercomputers, it is often necessary to invoke the executable indirectly, from what is known as a "job script". Such scripts typically use a (system-dependent) variation of one of the languages above. For an introduction to such languages see, for example, Robbins and Beebe [25]. Also in Windows there are several native technologies, such as Windows Script Host, the CMD shell, and Windows PowerShell (see, e.g., Knittel [13] for details). Although the shells may often seem to be primitive as programming languages, their distinctive advantage is that they seamlessly integrate with the rest of the system. On Unix in particular, a fundamental principle is to write programs that do a particular task well, and to design these so that they can easily communicate with one another, through streams of text[91] (see Raymond [24]). The shells were designed to fit into this picture as "glue"-languages, which make invocation of programs and pipelining of output easier.

Another class of scripted languages that can be used with Fortran are the more general-purpose R, Python, MATLAB, and octave. These offer some valuable tools, such as support for advanced statistics, visualization, and computational steering.

5.6.2 Software Libraries

Due to the long history of Fortran, it is natural that a large collection of programs and libraries has been created. Many of these are available, under open-source or commercial licenses, and are of high quality. Therefore, it makes sense to consider, when evaluating the requirements of a new application, if any of these libraries and programs could be used, to reduce the development costs. We provide a short overview here, for some of these libraries that are relevant to ESS. Note that, given the capabilities of modern Fortran to interoperate with C (as discussed in Sect. 5.4), it is also possible to use software libraries that were written in C.[92]

First of all, it is recommended to search for the desired functionality in the set of *intrinsic procedures* of Fortran. We could only cover a small subset of these here, so we refer to more advanced texts like Metcalf et al. [21] (especially *Chap. 8* and *Appendix A* therein) for a complete list.

Within the universe of software packages, an important role (especially for ESS) is occupied by *Linear Algebra* routines. The de facto standard library in this domain, especially for working with dense and banded matrices, is *Linear Algebra PACKage* (LAPACK). To ensure good performance on many platforms, this relies heavily on BLAS, which is a library for performing the lower-level computations. The latter is actually a collection of libraries, with a conventional interface—this allows hardware vendors to provide optimized versions of these libraries for their own platforms, such as Accelerate from Apple, *Core Math Library* (ACML) from AMD, *Engineering Scientific Soubroutine Library* (ESSL) from IBM, *Intel® Math Kernel Library* (MKL) from intel, etc. (consult the documentation of your system to see what is available). An alternative is *Automatically Tuned Linear Algebra Software* (ATLAS) [28], which can generate optimized versions of BLAS.

Categorized collections of Fortran libraries like netlib or *Guide to Available Mathematical Software* (GAMS) are good places to consult for other libraries.

---

[91] In particular, well-known tools for manipulating text are grep, awk, and sed.

[92] However, it may be necessary to write a thin "wrapper layer" of interface-blocks, based on the documentation of the C API of the library.

5.6.3 Visualization

Data visualization is very important in ESS. However, this is a vast field in itself, and we can only provide some pointers here. The issue of visualization relates to the hierarchical approach to I/O that we highlighted throughout the text, since choosing a suitable data format is a crucial prerequisite:

- ASCII files with minimal formatting are suitable for small and low-dimensional datasets (up to two independent coordinates). Most tools can operate on such files.
- netCDF-files are recommended (especially in ESS), since they are also widely supported, and tools like the *Climate Data Operators* (CDO) or the *netCDF Operators* (NCO) can be used for preprocessing the data files, when they would be too large to be comfortably used directly in the visualization software.

The concrete software package to use for visualization depends more on other factors, such as additional mathematical/graphics features that may be necessary:

- for simple visualizations, gnuplot is very suitable, especially for ASCII files
- interpreted languages, such as R, the *Generic Mapping Tools* (GMT) or MATLAB can be useful for more complex analyses, as they were either designed with ESS applications in mind (GMT), or accommodate a large set of packages for specialized functions (R, MATLAB); all of them can also operate on netCDF-data
- for 3D volume datasets, tools like *Parallel Visualization Application* (ParaView) (also supporting netCDF-files) can be used

5.6.4 Version Control

Software projects are, inherently, very dynamic: new features are added, parts of the code are restructured[93] for clarity or performance optimization, old bugs are fixed and, unfortunately, new bugs are introduced, etc. This process naturally leads to several versions, which can add management costs (for example, when trying to determine which change led to a certain bug).

To diminish these additional costs, version control systems were invented. They provide the concept of a "repository", which is where all revisions of the project are stored. As the project is evolving, developers can mark completion of certain milestones with "commits", usually accompanied by related comments. Any "committed" version can then be easily retrieved, and also compared against other versions (particularly useful). There is a lot of variation in the exact mechanics of these operations, and in the way special situations (like collaboration) are handled—these depend a lot on the type of system used. A rough classification of these systems identifies two

---

[93] Another common term for this is "refactoring".

classes,[94] either of which may be preferred, depending on the needs of the project
and on the background of the developers:

- *centralized systems* (e.g. `subversionn`): a central server is designated, where
  all project contributors upload their changes, and from which they get the latest
  revision of the code ("trunk"). With this hierarchy, it is always straightforward
  to locate the latest revision of the code. One limitation due to this server-client
  architecture is that network access becomes necessary for most operations.
- *distributed systems* ((`git`), `mercurial`, or `monotone`, etc.): with these systems, every developer commands a fully functional clone of the repository. This
  relaxes the constraint of constant network access, and also provides developers the
  means to test ideas in local "branches", before sharing them with others (often,
  such ideas are complex enough to benefit from version control on their own, which
  is easy to do in distributed systems). A possible disadvantage is that, since no two
  developers will make the same changes, the repositories can easily diverge in time,
  which may be a problem if a common version is desired (but these systems usually
  also provide excellent tools for synchronization).

5.6.5 Testing

Another technology that deserves some consideration is software testing systems.
The main idea here is to develop a large number of tests, which offer at least some
guarantee that the code (or some sub-components of it) are able to reproduce the
expected results for known inputs. Testing can be performed either at the system
level (for example, in ESS, a model's output can be tested for some setups that are
well documented in the literature), or at the function/module level ("unit tests"). The
latter are usually easier to implement, especially when supported through specialized
frameworks.[95] A key point is that these tests should be run regularly (ideally, every
time the software is recompiled – for example, when using `gmake`, a special target
can be created for this). This can dramatically reduce time spent chasing bugs when
changes are introduced, since a lower-level failure (as highlighted by the unit-testing
framework) is much easier to understand and fix than an error in the final model
results. Naturally, this system begins to pay off when a considerable number of unit
tests are written; a good rule of thumb is to write a unit test for every bug found in
the code.

A practice worth mentioning in this context is *test-driven development* (TDD),
which advocates writing of the unit test(s) *before* the corresponding feature is actually
implemented. The development cycle would progress along the following lines:

---

[94] The separation is not so clear-cut, since when a "centralized" system is used by a single developer
there is no need for a dedicated server; also, the "distributed" systems can be used in a server-client
fashion.

[95] See, for example, the *FORTRAN Unit Test Framework* (`fruit`).

1. add test for new feature
2. run the entire suite of unit tests, and confirm that the tests from previous step fail
3. add code for the new feature
4. run the test suite again, to confirm that the code passes all tests
5. repeat

The advantages of this approach are twofold: first, by constructing the unit tests, the developer obtains a more accurate picture of how the code for the new feature should behave; second, the tests provide immediate feedback (and, when they all pass, gratification) about the progress of the work.

The interested reader can find a discussion of unit testing in Fortran, in Markus [15].

References

1. Amdahl, G.M.: Validity of the single processor approach to achieving large scale computing capabilities. In: Proceedings of the 18–20 April 1967. Spring Joint Computer Conference, pp. 483–485. AFIPS'67 (Spring), ACM (1967)
2. Barakat, H.Z., Clark, J.A.: On the solution of the diffusion equations by numerical methods. J. Heat Transf. **88**(4), 421–427 (1966)
3. Calcote, J.: Autotools: A Practitioner's Guide to GNU Autoconf, Automake, and Libtool. No Starch Press, San Francisco (2010)
4. Caron, J.: On the suitability of BUFR and GRIB for archiving data. In: AGU Fall Meeting Abstracts, vol. 1, p. 1619 (2011)
5. Chandra, R., Dagum, L., Maydan, D., McDonald, J., Menon, R.: Parallel Programming in OpenMP. Morgen Kaufmann Publishers, San Francisco (2000)
6. Chapman, B., Jost, G.: Using OpenMP: Portable Shared Memory Parallel Programming. The MIT Press, Cambridge (2007)
7. Clerman, N.S., Spector, W.: Modern Fortran: Style and Usage. Cambridge University Press, Cambridge (2011)
8. Hager, G., Wellein, G.: Introduction to High Performance Computing for Scientists and Engineers. CRC Press, Boca Raton (2010)
9. Hao, W., Zhu, S.: Parallel iterative methods for parabolic equations. Int. J. Comput. Math. **86**(3), 431–440 (2009)
10. Hook, B.: Write Portable Code: An Introduction to Developing Software for Multiple Platforms. No Starch Press, San Francisco (2005)
11. Kerrisk, M.: The Linux Programming Interface: A Linux and UNIX System Programming Handbook. No Starch Press, San Francisco (2010)
12. Knight, S.: Building software with SCons. Comput. Sci. Eng. **7**(1), 79–88 (2005)
13. Knittel, B.: Windows 7 and Vista Guide to Scripting, Automation, and Command Line Tools. Que Publishing, Upper Saddle River (2010)
14. Locarnini, R.A., Mishonov, A.V., Antonov, J.I., Boyer, T.P., Garcia, H.E., Baranova, O.K., Zweng, M.M., Johnson, D.R.: World Ocean Atlas 2009 Volume 1: Temperature. In: Levitus, S. (ed.) NOAA Atlas NESDIS 68. U.S. Government Printing Office, Washington, D.C., p. 184 (2010), also available as http://www.nodc.noaa.gov/OC5/indprod.html
15. Markus, A.: Modern Fortran in Practice. Cambridge University Press, Cambridge (2012)
16. Martin, K., Hoffman, B.: Mastering CMake, 6th edn. Kitware Inc, New York (2013)
17. Martorell, X., Tallada, M., Duran, A., Balart, J., Ferrer, R., Ayguade, E., Labarta, J.: Techniques supporting threadprivate in OpenMP. In: Parallel and Distributed Processing Symposium, IPDPS 2006. 20th International, p. 7 (Apr 2006)

18. Mattson, T.G., Sanders, B.A., Massingill, B.: Patterns for Parallel Programming. Addison-Wesley Professional, Boston (2004)
19. McCool, M., Reinders, J., Robison, A.: Structured Parallel Programming: Patterns for Efficient Computation, 1st edn. Morgan Kaufmann, San Francisco (2012)
20. Mecklenburg, R.: Managing Projects with GNU Make (Nutshell Handbooks), 3rd edn. O'Reilly Media, Sebastopol (2004)
21. Metcalf, M., Reid, J., Cohen, M.: Modern Fortran Explained. Oxford University Press, Oxford (2011)
22. Moore, G.E.: Cramming more components onto integrated circuits. Electronics **38**(8), 114–117 (1965)
23. Overton, M.L.: Numerical Computing with IEEE Floating Point Arithmetic. Society for Industrial and Applied Mathematics, Philadelphia (2001)
24. Raymond, E.S.: The Art of UNIX Programming. Addison-Wesley Professional, Boston (2003)
25. Robbins, A., Beebe, N.: Classic Shell Scripting. O'Reilly Media, Sebastopol (2005)
26. Smith, P.: Software Build Systems: Principles and Experience. Addison-Wesley Professional, Boston (2011)
27. Sutter, H.: The free lunch is over: a fundamental turn toward concurrency in software. Dr. Dobb's J. **30**(3), 202–210 (2005)
28. Whaley, R.C., Petitet, A.: Minimizing development and maintenance costs in supporting persistently optimized BLAS. J. Softw. Pract. Exp. **35**(2), 101–121 (2005)

CPSIA information can be obtained
at www.ICGtesting.com
Printed in the USA
LVOW02s2233300617

539956LV00004B/9/P